W9-BNN-736

RADIOACTIVITY

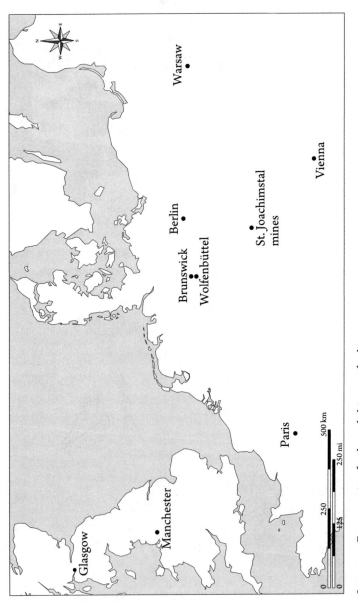

Important European sites for the early history of radioactivity.

RADIOACTIVITY

A History of a Mysterious Science

Marjorie C. Malley

OXFORD
UNIVERSITY PRESS

OXFORD
UNIVERSITY PRESS

Oxford University Press, Inc., publishes works that further Oxford University's
objective of excellence in research, scholarship, and education.

Oxford   New York
Auckland   Cape Town   Dar es Salaam   Hong Kong   Karachi
Kuala Lumpur   Madrid   Melbourne   Mexico City   Nairobi
New Delhi   Shanghai   Taipei   Toronto

With offices in
Argentina   Austria   Brazil   Chile   Czech Republic   France   Greece
Guatemala   Hungary   Italy   Japan   Poland   Portugal   Singapore
South Korea   Switzerland   Thailand   Turkey   Ukraine   Vietnam

Copyright © 2011 by Marjorie C. Malley

For titles covered by Section 112 of the US Higher Education Opportunity Act,
please visit www.oup.com/us/he for the latest information about pricing
and alternate formats.

Published by Oxford University Press, Inc.
198 Madison Avenue, New York, New York 10016
www.oup.com

Oxford is a registered trademark of Oxford University Press

Library of Congress Cataloging-in-Publication Data
Malley, Marjorie Caroline, 1941–
    Radioactivity : a history of a mysterious science / Marjorie C. Malley.
       p. cm.
    Includes bibliographical references and index.
      ISBN 978-0-19-976641-3
1. Radioactivity.   I. Title.
QC794.6.R3M35 2011
539.7′2—dc22   2010038979

9 8 7 6 5 4 3 2 1

Printed in the United States of America
on acid-free paper

*This book is dedicated
to all who wonder
and seek to understand*

CONTENTS

CONTENTS

PART TWO

MEASURING AND USING RADIOACTIVITY

PART THREE
BEYOND THE STORY

LIST OF ILLUSTRATIONS

MAP

FIGURES

APPENDICES

PREFACE

What would I recommend for her to read, a friend asked, for an overview of the history of radioactivity? Though I could think of many books and articles on different aspects of the field, I realized that nothing was available for my friend or other interested nonspecialists who did not have the background, time, or inclination to piece together a full historical narrative from the available materials.

I wrote this book to fill this need. It is a brief history of the science of radioactivity based on years of my research with published texts and manuscript materials supplemented with numerous secondary sources. It provides a broad and accurate history while avoiding excessive technical detail. This book is suitable for specialists in related fields such as physics, chemistry, and history; for nonspecialists wishing to know more about this remarkable episode in modern science; and for readers interested in the world at the turn of the twentieth century. Having spent years teaching and developing curricula, I am especially interested in making the history of science accessible for students and for teachers.

Radioactivity entered history at the onset of major transitions in science, and itself propelled these changes. The field's brief trajectory, spanning three decades, contrasts starkly with the interest it generated and the far-reaching results it produced. The story of radioactivity provides a window into the era's science and its cultural matrix. It illuminates the scientific process and the ongoing human quest for understanding.

Radioactivity has the dual attractions of a fascinating history and dramatic consequences for humanity. The atomic bomb, nuclear power, and changing relations of science to government and the military are obvious results. Though important in their own right, to concentrate on these outcomes would risk distorting the history of radioactivity by viewing the past through the eyes of the present. The radioactivity researchers worked in a very different environment from scientists several decades later. To them radioactivity was an enigma, a discovery with many possibilities for investigation, rather than a prelude to unknowable future developments. This work presents radioactivity's history as it unfolded at the time, a unique, exciting, and instructive story with a particular historical context.

From its modest beginnings as a minor phenomenon, radioactivity quickly developed into a major research field. Mysterious from the start, the new science was an intractable riddle until it was absorbed into emerging areas of physics. I have focused on this key feature of radioactivity, portraying the allure, challenge, and excitement of a totally unanticipated and mysterious phenomenon and the efforts of scientists to comprehend it.

The book's first section seeks to capture the puzzlement and suspense for radioactivity's pioneers by leading the reader through the twists and turns, surprises and dead ends which researchers experienced as they pursued their goal of understanding radioactivity. By following this story, the reader can better appreciate the processes involved in developing, testing, and revising scientific explanations.

Radioactivity's history includes applications, methods and instruments, and the institutional structures that supported the new science. These are reviewed in the second section of the book.

The history of the new science illustrates patterns and themes that have animated science and the universal human adventure. The final section of this book identifies and analyzes some factors which informed radioactivity's development and illustrate the ongoing human search to comprehend the world.

I am grateful to the many persons and institutions who supported my first researches on radioactivity many years ago, including the University of California at Berkeley and John Heilbron, who suggested radioactivity as a research topic and guided my first efforts. My late parents, Raymond and Alice Malley, inspired my interest in education. A remark by Jan Gillie led me to begin this book, and my late brother John Malley kept me on track. Tara Hornell made helpful suggestions on several chapters. David Haase, Physics Department at North Carolina State University, suggested improvements to Appendix 3.

I am especially indebted to Kristin Hornell for her detailed reading and critique of the text and for numerous valuable suggestions. Any errors remaining in this book are my responsibility.

I would like to thank editor Phyllis Cohen for her capable guidance and prompt responsiveness throughout the acquisition and editorial processes. The helpfulness of Hallie Stebbins, Jennifer Bossert, Jennifer Kowing, Woody Gilmartin, and others at Oxford University Press are much appreciated.

I would also like to thank the anonymous reviewers who commented on my book proposal.

I wish to acknowledge the institutions and individuals who provided illustrations and permissions and those who gave technical assistance, including Scott Prouty and the Emilio Segrè Visual Archives, American Institute of Physics; Thomas W. Buchler, CEO, and Waltraud Grove of Buchler GmbH, Brunswick, Germany; Adam Perkins, Don Manning, Ruth Long, and the Syndics of Cambridge University Library; Bailey Dabney and Randy Cowling, The Claremore Progress; Matthias Röschner and the Deutsches Museum; Rudolf Fricke, Wolfenbüttel, Germany; Nathalie Huchette, Natalie Pigéard, and the Musée Curie, Institut Curie; Paul Hogrian and the Library of Congress; Dianna Everett, Oklahoma Historical Society; JoAnn Palmeri, Kerry Magruder, and The History of Science Collections, University of Oklahoma Libraries; Professor C. M. R. Fowler and Ernest Rutherford's family; David N. Hall and The Frederick Soddy Trust; Regina Frey, Chemistry Department, Washington University at St. Louis; Peter Graf and the Zentralbibliothek für Physik in Vienna; Dennis and Donna Grossman, James Hornell, and Earl W. Hall. Many thanks to all.

I am extremely grateful to my husband, James Hornell, for his constant support, encouragement, suggestions, photography, computer troubleshooting, and general helpfulness throughout the lengthy process of completing this book.

Unless otherwise indicated, translations are by the author.

INTRODUCTION

Radioactivity burst into the world without warning. No precursors foreshadowed it, and nothing in nineteenth-century physics could have predicted it. Barely noticed at first, within a few years radioactivity became a prime topic for researchers. For the public, radioactivity transformed from a minor curiosity into a potential font of miracles.

The 1896 discovery of invisible uranium rays dramatically changed physics and chemistry as well as the lives of future generations. Fields as diverse as geology, archaeology, biology, medicine, meteorology, philosophy, power generation, and warfare were altered by the new findings. Radioactivity revealed details of the structure of matter and provided tools for investigating it. It revolutionized ideas about energy, eventually ushering in nuclear physics and a new era.

Radioactivity furnished evidence for probability's reign within the atom, influencing ideas about causality's place in the universe. It altered theories of the age of the earth and supplied new methods for dating artifacts. It inspired innovations ranging from new scientific instruments and techniques to smoke detectors and luminous watch dials. In chemistry, radioactivity unlocked the puzzling periodic table of the elements, altering ideas about elements and expanding the table itself.

Radioactivity provided novel ways for treating cancer and for understanding physiological processes. It fueled new enterprises to find and use radioactive materials for medical, industrial, and commercial purposes. The search for radioactive materials revealed their wide distribution and led to the discovery of cosmic rays.

The new science's political and social ramifications were far-reaching. Several national research institutes and special laboratories were created to study radioactivity, accelerating the concentration of scientific resources, scientific teamwork, and government influence on research. The new field fostered increased participation of women in physics. Radioactivity introduced unexpected health hazards for those who worked with it, whether they were academic researchers, miners, therapists, or factory workers. These problems spawned new organizations and regulatory agencies and contributed to a popular distrust and disillusionment with science that increased later in the century.

During the late 1920s radioactivity transmuted into nuclear physics. The fields of particle physics and nuclear chemistry also have roots in the 1896 discovery. Centralization of resources, government involvement, and political events intertwined with nuclear physics to create a matrix for nuclear weapons and nuclear reactors, which have forever changed the world.

Radioactivity was one of a series of surprises which greeted scientists at the turn of the century. It followed on the heels of the 1895 discovery of x rays. The previous year a gas had been found which chemists were unable to combine with other elements. Named *argon*, meaning "idle," it belonged to an entirely new family of chemical elements, the inert gases. In 1897 a Dutch physicist detected what became known as the *Zeeman effect*, an esoteric but theoretically significant action of magnetism on spectra which offered clues to atomic structure. In the same year the first subatomic particle, the electron, officially entered physics. The energy quantum debuted in 1900, special relativity theory in 1905, and general relativity in 1916. These innovations revolutionized physics during the next few decades. Applications of the new physics transformed society, while that milieu in turn altered the scientific enterprise.

The story of radioactivity is interwoven with the history of modern physics. This book tells radioactivity's story in that context, from the first discovery through subsequent startling findings and the questions and problems of researchers who grappled with this mysterious phenomenon. It outlines the rise of related industries, research institutions, and medical developments. It provides background on some contributors to the field and analyzes factors affecting radioactivity's growth and development. Finally, this book reflects on radioactivity's links to perennial human questions, quests, motifs, and themes.

PART ONE

A NEW SCIENCE

There are more things on heaven and earth …
Than are dreamt of in your philosophy.
　　　　　　　　　　　—Shakespeare, Hamlet, act 1, scene 5

The turn of the twentieth century was full of surprises for physicists.
Amid the unexpected developments, a new science blossomed. By 1919
this field had matured and physics was changing dramatically, as was
the world outside the laboratory.

The Beginnings

The Roentgen Rays, the Roentgen Rays,
What is this craze?

—*Photography*, 1896

THE SETTING

It was 1895 in Europe. France was embroiled in the notorious Dreyfus case, a court-martial tainted with anti-Semitism. England, France, and Italy were staking out claims to parts of Africa, while across the Atlantic pioneers streamed into the remaining American Indian territories. The remnants of the ancient Roman Empire, now reduced to the Austro-Hungarian Empire, simmered with tribal enmities and socioeconomic tensions among a mixture of ethnic groups.

A romantic strand was woven into the era's culture. Fashion favored flowing skirts, lace, and elaborate hats. Art Nouveau was all the rage, with lamps that resembled trees and other creations embodying stylized organic forms, while the Russian composer Tchaikovsky's *Swan Lake* premiered in St. Petersburg. Genteel ladies and dignified and learned men in top hats frequented elaborate theatrical productions as well as the theaters of the supernatural known as séances. Doomsayers and soothsayers and assorted fringe elements anticipated the change of centuries with escalating excitement.

If one listened, among the prophets and cranks who watched for the turn of the century could be heard an undercurrent of wild expectations, murmurs of matter transforming into energy, atoms reduced to vibrations in the ether, the ebb and flow of reality always changing yet always the same. This was the sound of the river of the ancient Greek philosopher Heraclitus, who taught that reality was like a river, forever changing yet forever the same. One never, according to Heraclitus, steps into the same river twice. Nothing is permanent but change. Now, in the late nineteenth century, Heraclitus' philosophy reappeared in modern popular guise, as his river metaphorically flowed through electromagnetic theory and carried along bits and pieces of scientific detritus into the teeming imagination of the expectant public.

In the biological realm, Charles Darwin's theory of evolution afforded a scheme which made change integral not only to living things, but to the forms of life themselves. Evolutionary models, some predating Darwin, were applied to the earth, the solar system, and the periodic table of the chemical elements, as well as to studies of culture, society, and politics. Later, the transmutation theory, which claimed that some elements could change into other elements, would suggest that radioactivity could be assimilated to this theme.

In physics, electricity and the beautiful and mysterious effects which it created in high voltage vacuum tubes were popular fields for investigation. Nineteenth-century advances in technology had made it possible to study matter and electricity in high vacuum. Three German instrument makers, Johann H. W. Geissler, Heinrich Rühmkorff, and Hermann J. P. Sprengel, revolutionized the study of electricity in rarefied gases.

In 1855 Geissler developed a pump which used mercury instead of pistons to remove air from containers. He also devised thin glass tubes with two metal pieces (electrodes) embedded in the glass that could be connected to an electrical generator. The electrodes would receive opposite electrical charges, which by convention scientists labeled "negative" and "positive." Depending on the voltage applied from the generator and the amount of air still in the tube, magical patterns of dancing light and darkness would appear in the tubes. These patterns, called "vacuum discharges," signaled the transfer, or discharge, of electricity between the oppositely charged electrodes.

Rühmkorff improved the devices (induction coils) that could generate the high voltages needed to create discharges (Figure 1-1),

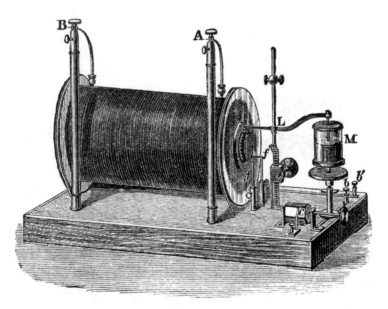

Figure 1-1. Rühmkorff coil. This device is a transformer for producing high voltages. From Silvanus P. Thompson, *Elementary Lessons in Electricity & Magnetism* (New York: Macmillan, 1901), 219.

while in 1865 Sprengel modified Geissler's pump. Sprengel's pump could remove nearly all of the air from glass tubes designed for this purpose, called "vacuum tubes." The efficient pump facilitated study of the vacuum discharges. These striking visual displays fascinated physicists, who tried to make sense of their complexity by grouping the displays into categories and finding the conditions needed to produce each type (Figure 1-2). Some even imagined replacing the concept of matter with evanescent energy forms like the luminous electrical discharges. The energy forms supposedly disappeared into and reappeared from a ghostlike invisible fluid called the "universal ether."

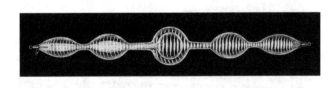

Figure 1-2. Geissler tube. From Henry E. Roscoe, *Spectrum Analysis* (London: Macmillan, 1869), 106.

The idea of an invisible fluid which pervaded the universe was not new, but the British physicist James Clerk Maxwell had made it central to a remarkable mathematical-physical theory. Maxwell's theory joined the invisible forces of electricity and magnetism into a complex known as *electromagnetism*, and linked electromagnetism with light.

Maxwell believed that light was a wave motion in a weightless fluid of electromagnetic origin, the ether. He predicted another type of wave motion in the ether, later known as radio waves. Light, radio waves, and other related wave motions became known as *electromagnetic radiations.*

Some scientists speculated that matter might be made of ether particles. The famed Russian chemist Dmitri Mendeleyev included ether in his revised (1903) periodic table of the elements. A vocal minority promulgated energetics, a scientific philosophy which viewed energy as primary, relegating the world of our senses, of sights and sounds and smells and myriads of discrete objects, to a phantom existence as mere energy forms. The British mathematician James Jeans envisioned oppositely charged particles annihilating each other in a burst of energy, presaging later ideas about antimatter.

Matter, the stuff we can see and touch, had long been viewed as the primary reality. Energy was only a manifestation of matter in motion, whether of invisible particles, living beings, or machines. Maxwell's theory made it possible to compute equivalences between matter and energy, implying they could be converted into one another. From there it was a simple step to stand the traditional view on its head and make matter into a form of energy. Perhaps material objects were merely disturbances in a sea of ether. Most scientists did not subscribe to this extreme view; but its adherents were outspoken, and it was taken up by some members of the general public.

The invisible force fields of electromagnetism, mysterious electrical discharges, the invisible ether, and discoveries of invisible light and other radiations reinforced speculations about an unseen ghost world that could be contacted by sensitive intermediaries, or mediums. These were not fringe ideas. In the United States and Europe many educated persons, including prominent scientists, tried to contact deceased loved ones in sessions (or séances in the preferred French) which mediums held in darkened rooms. The popularity of what was known as spiritualism towards the end of the nineteenth century led to serious studies of this idea as well as concerted efforts by scientists and official commissions to expose the mediums as frauds.

RAYS AND RADIATION

Both the interest in spiritualism and speculations about conversion of matter and energy were boosted by discoveries of new rays. The term *ray*, used loosely for a beam of light, expanded during the nineteenth century to include invisible forms of light, after investigators found light-like effects occurring beyond the edges of the visible spectrum. Light beyond the violet edge became known as ultraviolet light (from *ultra*, meaning "beyond") and light below the red edge became infrared light (from *infra*, meaning "below"). The terms *ray* and *radiation* were used interchangeably during this period and beyond.

Later, a variety of poorly understood electrical, chemical, and photographic effects were attributed to invisible rays. In addition to infrared and ultraviolet light, the turn-of-the-century observer might encounter phosphorus light, light from fireflies, Le Bon's black light, diacathodic and paracathodic rays,

Lenard's rays, Wiedemann's discharge rays, Goldstein's canal rays, and more.

Most studied were the cathode rays, so named because they came from the negatively charged electrode (cathode) of a vacuum tube. Though invisible, cathode rays made fluorescent materials glow. To study these rays, researchers used vacuum tubes coated at one end with a fluorescent material (Figure 1-3). Magnets changed the paths of cathode rays, deflecting them as though they were a beam of negatively charged particles. Yet, unlike any known form of matter, these rays could pass through metal foil. Perhaps cathode rays were a new, rarefied form of matter, neither solid, liquid, nor gas—a fourth state of matter, according to British chemist Sir William Crookes.

An independent researcher and consultant to industry, Crookes maintained a laboratory in his home and founded the journal *The*

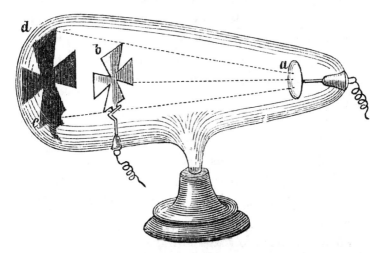

Figure 1-3. Cathode ray tube. Electrons travel from the cathode (a) towards the anode (b). Rays that strike the glass tube behind the anode cause a phosphorescent coating to glow. The cross-shaped anode in this tube is designed to cast a shadow, demonstrating that the rays travel in straight lines. From Eugen von Lommel, *Lehrbuch der Experimentalphysik* (Leipzig: J. A. Barth, 1895), 343.

Chemical News. His many important researches included extensive work with cathode rays. Many scientists, especially in Britain, agreed with Crookes that the rays must be some type of particle. Other scientists, particularly in Germany, thought the cathode rays were a kind of invisible light. They believed that only a wavelike process like light could be transmitted through foil.

Into this surreal atmosphere, just at the end of 1895, came an announcement which electrified a world already fascinated by rays. Wilhelm Röntgen, professor of physics at Würzburg University in Bavaria, had found a new type of invisible radiation which could penetrate solid, opaque objects!

Röntgen's rays came from vacuum tubes designed to study cathode rays. He called this radiation "X" because nothing was known about it.

Röntgen had been investigating reports that cathode rays could pass through aluminum foil. He used various tubes to produce the rays, including one developed by his countryman Philipp Lenard. The tubes had an aluminum "window" at one end to transmit the cathode rays. First he would need to eliminate all light from the laboratory. After making the room completely dark, he turned on the apparatus to ensure that it was working properly.

To his surprise, a screen made with fluorescent material, resting some distance from the apparatus, glowed whenever he switched on the tube. Röntgen had not expected anything like this to happen because he had covered his tube with black cardboard to keep any light from escaping, and in any case cathode rays could not travel that far in air.

Röntgen decided to track down the reason for the screen's behavior. After days of feverish work, his persistence brought momentous results. He had discovered an entirely new radiation that was much more penetrating than anyone had imagined. These rays caused air to conduct electricity. Unlike the cathode rays, they were not affected by a magnet. To demonstrate his discovery, Röntgen published the first photographic image of the bones in a human hand, courtesy of Frau Röntgen. Other scientists followed suit. Once journalists got hold of x-ray photographs, the public went wild, and the quiet, reserved professor became instantly famous.

Later, reports circulated of physicists narrowly missing the discovery of x rays. When some physicists noticed that photographic plates stored near cathode ray tubes became fogged, they simply moved their plates. Crookes even returned damaged plates to the manufacturer. Röntgen's colleague Philipp Lenard performed extensive experiments with his special cathode ray tubes, but he did not realize that cathode rays produced a new radiation that caused some of his results.

The medical implications of the discovery were stunning. To be able to see inside the body was a wondrous advance. Physicians quickly began using the rays to visualize fractures and locate foreign bodies in their patients. Soon, they predicted, x rays would allow them to visualize internal organs and find tumors. Perhaps vivisection would become obsolete.

The discovery provoked dubious claims and enterprises. A farmer reported that he had used x rays to turn a common metal into gold, and a Frenchman claimed to have photographed the soul. Spiritualists hoped the rays would enhance their séances. Some in the general public feared a loss of privacy and modesty— could the rays visually disrobe a person?

Taken aback by the popular reactions and unwilling to sacrifice his precious time, Röntgen shied from publicity and continued to investigate the rays. In 1901 he received the first Nobel Prize for Physics for his discovery. These prestigious prizes were instituted by Alfred B. Nobel, a wealthy Swedish industrialist and the inventor of dynamite, to honor major contributors to physics, chemistry, physiology or medicine, literature, and peace.

The discovery of x rays prompted another line of research. Could these new rays have something to do with the light that cathode rays produced when they struck a screen coated with a fluorescent mineral? Perhaps other fluorescent materials gave off x rays. This idea must have occurred to many physicists. After attending a presentation on Röntgen rays (x rays were often called "Röntgen rays") in January 1896 by the philosopher and mathematician Henri Poincaré, a French physicist decided to test the hypothesis that x rays were connected with fluorescence.

BECQUEREL'S DISCOVERY

Antoine-Henri Becquerel, the son and grandson of eminent French physicists, was the right person living at the right place and time (Figure 1-4). As director of Paris's Museum of Natural History, he was in charge of a large collection of luminescent minerals that his father had assembled. When these minerals absorbed light, they would emit light of wavelengths (colors) different from the original source. If the luminescence disappeared when the incident light was removed, the phenomenon was often called "fluorescence." If the luminescence continued, it was usually called "phosphorescence." Optical luminescence was Edmond Becquerel's life work, and his son Henri had also made a name for himself with optical research. Not only did Henri Becquerel have a distinguished pedigree, he even held the position formerly occupied by his father, and earlier by his grandfather Antoine-César Becquerel.

Figure 1-4. Antoine-Henri Becquerel. From the Generalstabens Litograficka Anstalt (Sweden), courtesy of the Library of Congress.

Back in the museum, Becquerel began testing samples from his father's collection. He was particularly interested in a luminescent uranium mineral. Named in 1789 after the newly discovered planet Uranus, uranium was a heavy metal found mainly in central European mines. It was used to color ceramics and glass. There had been no sign that anything was special about it.

Yet, Becquerel had grounds for focusing on uranium minerals. He believed a heavy mineral would be most suitable for converting visible light into x rays. He may have suspected a conversion might be fostered by some peculiarities in uranium's spectrum. Further, Becquerel's father had noticed that uranium minerals produced especially bright phosphorescence, and his countryman Abel Niepce de Saint Victor had noticed a persistent photographic effect with some uranium salts.

In 1896 photography commonly employed glass plates coated with a light-sensitive substance. Becquerel placed the uranium mineral on a photographic plate covered by black paper, then exposed it to sunlight in order to make it fluoresce. The black paper would keep out visible light, but it would not block x rays.

After allowing time for an x-ray image to form, Becquerel developed the plate. When he saw a distinct image of the mineral appear, he was not too surprised, since he had seen reports of invisible rays being emitted during phosphorescence. He tried the experiment on another day; but the sun was mostly hidden by clouds, so he brought his apparatus inside and stored it in a drawer. After waiting futilely for better weather, Becquerel developed the plate, expecting to find only a faint impression from its brief exposure to daylight. To his astonishment, a very intense image of the sample appeared!

This was puzzling, because phosphorescent and fluorescent minerals normally will not glow if they have not been exposed to light. Perhaps the mineral had managed to absorb enough light on the cloudy day to make a distinct image. But when Becquerel tried his next experiment, this time carefully shielding the sample from light, it still darkened the plate! Apparently this uranium mineral could phosphoresce for an unusually long time.

In England, the electrical engineer Silvanus P. Thompson had just found that a uranium compound gave off invisible rays long after it had been exposed to light or electricity. Thompson named this property "hyperphosphorescence," but when he learned that Becquerel had published similar results, he discontinued his research. Becquerel pursued his unusual findings, believing that the effect would fade if he waited long enough. Keeping his uranium compound in darkness, he continued to develop photographic plates and compare the images imprinted on them by the uranium.

Hours turned into days, then weeks, then months; yet, even after more than a year, uranium's power had hardly abated. Apparently, light was not required for the invisible rays to emerge. What the rays did seem to need was uranium. Everything Becquerel tested which contained this element darkened a photographic plate covered with black paper, while (with a few exceptions, later shown to be caused by experimental errors) other minerals did not. Even uranium minerals that were not phosphorescent imprinted images through the paper. If phosphorescence was producing the uranium rays, it would have to be a very different sort from the phosphorescence Becquerel and his father had studied.

Still, Becquerel held to his phosphorescence hypothesis. He tried unsuccessfully to destroy uranium's activity by dissolving and recrystallizing a uranium salt in darkness, a procedure known to destroy phosphorescence in other materials. Nothing he did stopped uranium from sending out invisible rays. When he found that uranium metal produced even stronger effects than uranium compounds, Becquerel did not change his views, even though metals were not supposed to be able to fluoresce. Instead, he concluded that uranium was an exception to this rule. The invisible rays, he decided, must come from the element uranium. In hindsight, this was a very important deduction, but Becquerel attached no special significance to it.

Testing to compare his newly discovered rays to other radiations, Becquerel found that uranium's rays electrified air and passed through cardboard, aluminum, copper, and platinum. Ultraviolet light, cathode rays, and x rays could all make air conduct electricity. In contrast, only x rays readily penetrated opaque materials.[1] This ability suggested that uranium's rays were a type of x ray. Still, Becquerel believed that phosphorescence caused uranium's strange behavior, which meant the uranium rays should be a form of light.

Becquerel drew on his experience to test uranium rays for their optical properties. Light's trademark properties were reflection, refraction, and polarization, and by the end of March 1896 Becquerel claimed to have detected all three in his uranium rays. (Later these researches were found to be faulty.) Further distinguishing his rays from Röntgen's, Becquerel showed that uranium rays and x rays were absorbed differently.

In 1897 a new discovery captured Becquerel's attention. For years he had searched for evidence that magnetism could influence light, something predicted decades earlier by the brilliant researcher Michael Faraday. Now the proof was at hand. A Dutch physicist at the University of Leiden, Pieter Zeeman, reported that powerful magnets could break a single spectral line into three lines. Having wrapped up his work on the uranium rays and thereby insuring his priority, Becquerel eagerly returned to his old interest. He spent the next year and a half investigating what later became known as the Zeeman effect.

Becquerel's flagging interest in the rays from uranium matched the verdict of the larger scientific community, which considered them a curiosity of no great significance. Because Becquerel's rays could penetrate matter, most believed (in contrast to Becquerel) they were a type of x ray. The Röntgen rays swallowed up Becquerel's rays and became the era's hot topic, with dozens of scientists taking up their study.

Researchers hoped to find knowledge and fame by mimicking Röntgen's discovery. In their haste to publish, some neglected basic experimental precautions and controls. Since other agents readily affected equipment and materials used to search for invisible rays, errors were rampant. For instance, the French physician-author

Gustave Le Bon, best remembered for his work on the psychology of crowds, wrote prolifically on "black light," a spurious invisible radiation. He believed that radioactivity was part of a general disintegration of all matter. A few years later René Blondel's "N-rays" (named after Nancy, France, where Blondel supposedly discovered them) set off a sensation in France, until they too were discredited.

Le Bon's work was symptomatic of a contemporary illusion in which the world seemed to be full of mysterious entities in a state of flux. Perhaps the fascination with the invisible world of spiritualism and the occult, especially in France and Britain, enhanced people's willingness to believe in unproven invisible rays. In turn, the discoveries of new rays were used as grist for all sorts of pseudoscientific speculation. This was the atmosphere for the next surprising turn of events.

The Curies

She has a very decided look, not to say stubborn.

> —*Jacques Curie commenting on his brother*
> *Pierre's photo of Maria Skłodowska*

It's pretty hard, this life that we have chosen.

> —*Pierre Curie*

MARIA SKŁODOWSKA

In 1894, a young Polish student in Paris was looking for a thesis research topic. She intended to earn a Ph.D. in physics—something no woman had achieved there. But Maria Skłodowska would never allow tradition to deflect her from her goals. Once decided, she was undauntable.

Born in Warsaw in 1867 to educated parents of limited financial means, Maria Skłodowska showed early signs of intellectual giftedness and remarkable powers of concentration. The youngest of five children, she became an avid reader and an outstanding student who eagerly consumed all that the world of learning had to offer.

The early death of her eldest sister, followed by the mother she adored, left a lasting imprint on Maria's young psyche and predisposed her to bouts of depression. Madame Skłodowska had been deeply religious, and Maria had followed her lead, with the religious upbringing shaping Maria's outlook and values. But after her mother's death Maria Skłodowska's religious faith began to die as well, for how could a loving God allow such a cruel thing to happen?

While growing up in Russian-occupied Poland, Maria imbibed the spirit of rebellion and ardent patriotism of her oppressed people. As a teenager she took part in an underground university, and later risked imprisonment to teach Polish peasant children to read and write in their native tongue. Maria adopted a popular philosophy called positivism, which professed the perfectibility of humanity and stressed education. She learned about socialism, another movement that promoted gender equality and aimed to improve society. This was not a young woman with whom one would trifle.

The elder Skłodowskis were both teachers, and assumed their children would earn higher degrees. Since she enjoyed and excelled in so many subjects, Maria was not sure what to pursue. She was interested in sociology and particularly loved literature. Finally, she resolved to study science and mathematics. The problem was, How? It was impossible to pursue her dream in Poland, for the universities would not admit women, and her family had little money to support her in any event. France, with its liberal tradition and traditional bonds with Catholic countries, attracted many young Poles. Maria decided to study in Paris. She and her elder sister Bronya, who wished to study medicine, agreed to take turns supporting one other.

As the elder, Bronya began her studies first, while Maria spent several years tutoring students and working as a governess. It seemed that Maria's savings would never be enough to support her education. Finally Bronya, who had married in the meantime, was able to convince her sister to come to Paris and live with her.

It was hard for Maria to leave her homeland, and especially to leave her father. Promising him that she would return to follow a teaching career in Poland, Maria took the long train journey to Paris. She enrolled at the University of Paris (the Sorbonne) in the fall of 1891, using the French form of her name, *Marie*.

The busy household of Bronya and her husband Casimir Dłuski made it difficult for Marie to concentrate. Further, the home was not close to the Sorbonne. After some months Marie moved into her own quarters nearer to the university. There she took on a sparse, monastic existence, necessitated by poverty but agreeable to her own inner desires. Her studies, seasoned with social interludes with the expatriate Polish community, became her existence.

As so often happens with those who give up the outward practice of their religion, the inner forms remain, indelible marks upon character. While professing a tolerant agnosticism, Marie Skłodowska took on the behavior of one dedicated to the religious life. Her garret room became her monastic cell; her studies became her devotion. She increasingly clothed herself in plain, simple garments, preferably black, the color of self-denial and the prescribed uniform of clerics and nuns. "Peace and contemplation," she later remarked, were "the true atmosphere of a laboratory," with laboratories already designated as "the temples of the future" by French chemist Louis Pasteur.[1]

In place of existential truth and moral sanctity, Marie sought scientific truth and scientific probity. She refused to attack religion, admittedly envying those who found faith easy. Throughout her life her inner stance remained religious to the core, with the pursuit of science replacing the traditional goals of religion and the laboratory replacing the sanctuary.

The diligence of the young secular nun was fruitful. Marie Skłodowska earned two master's degrees, receiving the top grade for the physics examination in 1893 and earning second place in the mathematics examination the following year.

A CONSEQUENTIAL MEETING

Marie then began a study of the magnetic properties of different kinds of steel for the Society for the Encouragement of National Industry. As she needed a place to perform the experiments, one of her countrymen introduced her to a French physicist who might have a room available at the Municipal School of Physics and Applied Chemistry. His name was Pierre Curie.

The son and grandson of physicians, Curie was an idealistic, independent thinker who had adopted his father's liberal and free-thinking views. His father was a researcher at heart and had published works on tuberculosis, while his maternal grandfather and uncles had developed inventions for dye and cloth manufacture. It was natural for Pierre Curie to develop scientific interests. After receiving degrees in physics and mathematics, he pursued original researches on crystal symmetry and on electrical properties of crystals.

A dreamy introvert, Pierre Curie was not interested in status or idle conversation. Later, Marie would remember being "struck by the open expression on his face and by the slight suggestion of detachment in his whole attitude."[2]

Marie and Pierre found they had much in common, and their friendship grew and deepened. Marie was conflicted between her love for her native Poland and her attraction to Pierre, for she had intended to settle in Poland after her studies, yet did not believe it would be fair to ask Pierre to leave France. Finally the decision was made, and the two physicists married in Paris in 1895. The next year Marie passed the examination which qualified her to teach. In 1897 the Curie's first child, Irène, was born about the time that Marie completed her researches on steel (Figure 2-1).

There was never any question of Marie leaving research once she became a mother. Both she and Pierre were committed to research, and the only real issue was how the necessary accommodations would be made. The couple hired a servant, and Pierre's newly widowed father came to live with them. Dr. Eugène Curie became his grandchild's devoted companion.

Marie's next goal was the doctorate degree. Intrigued by Henri Becquerel's reports on uranium rays and wishing to pursue a topic which had not been studied extensively, she decided to investigate these rays for her dissertation.

Figure 2-1. Marie Curie. Courtesy of the AIP Emilio Segrè Visual Archives, W. F. Meggers Gallery of Nobel Laureates.

Becquerel had used photography to investigate uranium rays, a medium which can produce striking images but is difficult to quantify. Marie decided instead to track the invisible rays through their electrical effects, a decision with far-reaching consequences.

At that time instruments called *electroscopes* were used to detect electrical effects in the air. Scientists recognized that these electrical effects were caused by moving charged particles they called "ions," from a Greek word meaning "traveler." The simplest electroscope contains two thin pieces of metal foil attached to a metal rod that is suspended from an insulator. The apparatus can be shielded from stray electrical effects by a metal housing. If an electrified, or "charged" body is touched to the rod, the electrical charge will pass to the electroscope leaves and cause them to separate (Figure 2-2).

Figure 2-2. Gold leaf electroscope. From Silvanus P. Thompson, *Elementary Lessons in Electricity & Magnetism* (New York: Macmillan, 1901), 16.

When radiations capable of breaking molecules apart into ions (called "ionizing radiations") bombard the air molecules near the electroscope, the air will become a conductor and carry away some (or all) of the electroscope's charge. The electroscope leaves then fall towards each other, since less (or no) charge is available to counteract the pull of gravity. The experimenter can determine the strength of the radiation by the change in angle of the electroscope's leaves.

Some years earlier Pierre Curie and his brother Jacques had found that quartz crystals gave off electrical signals when they were compressed. Pierre used this property (known as the *piezoelectric effect*) to devise an unusually sensitive electroscope, the quartz piezoelectroscope (Figure 2-3). This instrument compares electrical effects which radiations produce on the crystal to the effects produced by known weights. By using the quartz piezoelectroscope, Marie would be able to detect very small differences in ionizing radiation. She could then compare the strengths of invisible rays emitted by different substances.

The director of Pierre's school found a small storeroom where Marie could do her experiments. She tested uranium compounds and minerals loaned to her by other scientists. She soon found that each substance's power of emitting ionizing rays (which she called "activity") depended directly upon the amount of uranium it contained, rather than on its physical or chemical state. The activity seemed to belong to the element uranium itself, as Becquerel had concluded.

Might other elements also give off ionizing rays? Marie borrowed samples of the other elements for testing. Materials containing the rare metal thorium made her electroscope lose its charge. First identified in a mineral from Norway, thorium had been named in 1829 after the Norse god Thor. Apparently, thorium

Figure 2-3. Pierre Curie with the electroscope he invented. From Marie Curie, *Pierre Curie*, trans. Charlotte and Vernon Kellog (New York: Macmillan, 1923). Image courtesy of History of Science Collections, University of Oklahoma Libraries.

also emitted ionizing rays. But a German chemist, Gerhard Carl Schmidt, had already published that finding.

Of all the elements Curie tested, only two—uranium and thorium—gave off invisible ionizing rays, a power which the Curies named "radio-activity" in 1898. These rays became generally known as "Becquerel rays," a term first used by the Curies in the same year.

If radioactivity was a property of certain elements regardless of their physical or chemical state, radioactivity must be a property of the atoms of these elements, an atomic property. At that time it was considered very important to distinguish between atomic properties and molecular properties. Atomic properties were presumed to be unchanging characteristics of individual atoms, while molecular properties characterized combinations of atoms, such as chemical compounds. As an atomic property, radioactivity would take its place among the established atomic properties of weight, spectra, and valence.

Looking back, it is tempting to read more into the term *atomic property* than it meant at the time, and Marie Curie herself encouraged this extrapolation. Becquerel had already concluded that radioactivity was a property of a specific element. Curie went one step further by stating that radioactivity was an atomic property. This insight was significant. However, the term *atomic* in 1898 did not have the associations it acquired after the discovery of atomic transmutation, especially after atomic reactors and bombs entered the picture.

NEW ELEMENTS!

Marie decided to investigate an intriguing result from her survey of minerals. Since the quartz piezoelectroscope allowed her to assign precise numerical values to each substance's activity, she had noticed something that photography could not make evident: Two minerals, pitchblende and chalcolite, showed much more activity than their content of uranium or thorium warranted. On the other hand, artificially prepared chalcolite showed no unusual activity. Since she had tested all the known elements, only one possibility remained. The natural minerals must contain a new, highly active element!

Finding the hypothetical new element would require lengthy chemical separations. Marie would attack the sample with a series of reagents, each time collecting the soluble and insoluble portions and subjecting them to new rounds of tests. To work quickly Marie would need help, and Pierre, excited about the direction his wife's work was taking, decided to join her. Because chemical expertise would be important, Pierre recruited his colleague Gustave Bémont.

They began with about 3.5 ounces of pitchblende. The Curies expected to find no more than one part in one hundred of the new element in the ore. Perhaps it was best they did not realize the proportion would be closer to one in a million.

By July 1898 the Curies realized that their chemical separations were concentrating radioactivity in the insoluble portions which contained the element bismuth. Since bismuth was not radioactive, the activity must come from another, unknown substance which was chemically related to it. This was their new element! Marie, ever mindful of her beloved homeland, called it "polonium." But this was not to be the last word on pitchblende, for in a few months they had evidence for a second new element. Different chemical treatments, reactions which separate elements that behave like barium, also concentrated the radioactivity. For this new element they chose the name "radium" (Figure 2-4).

Unknown to the Curies, a German chemist had also found a new active body different from polonium. A spectroscopic test identified the heavy metal barium, but since barium is not radioactive, another ingredient must have caused the radioactivity. After reading the Curies' papers, Friedrich Giesel realized that he and the Curies were investigating the same substance. Giesel's preparation glowed spontaneously, giving off (he later remarked) the "most splendid" bluish light, so bright he could read by it.[3]

Giesel wrote to the Curies about his results, explaining that he could concentrate the new material more quickly by crystallizing it from bromide salts than was possible with the chlorides the

Figure 2-4. Extraction of radium in the old shed. From Marie Curie, *Pierre Curie*, trans. Charlotte and Vernon Kellog (New York: Macmillan, 1923). Image courtesy of History of Science Collections, University of Oklahoma Libraries.

Curies had used. Giesel used his technique to produce radium at Buchler & Company's chemical factory in Brunswick, Germany, making his employer the major supplier for European scientists for several years (Figure 2-5). Giesel was the first person to apply a standard chemical test to his new substance. He ignited a small quantity in the flame of a Bunsen burner. Unlike barium, which produces a green flame, Giesel's material colored the flame a beautiful carmine red.[4]

The Curies' findings caught the attention of the scientific world. The discovery of new elements was always a momentous occasion. But since the Curies had used a totally new method to find polonium and radium, an electrical method, their discovery was especially remarkable—and controversial. Only recently had the chemical community grudgingly accepted *spectroscopy*, a method

Figure 2-5. Friedrich Giesel. Courtesy of Mr. Thomas W. Buchler, Buchler GmbH.

for analyzing light by breaking it into a rainbow of colors, as a legitimate tool for chemists. Several decades earlier, when the British scientists Norman Lockyer and Edward Frankland announced that sunlight, when passed through a spectroscope, revealed a new element in the sun—named "helium" after the Greek word for sun, *helios*—an unbelieving chemical community, long used to relying upon concrete sights and smells for its detective work, resisted this intrusion of physics with its abstractions into their field (Figure 2-6). Now chemists were being asked to accept the existence of invisible, imponderable elements on the basis of a physical device which supposedly recorded invisible rays! This was too much for some to believe.

Their skepticism was understandable, even if not always well informed. Chemical analysis is fraught with difficulties.

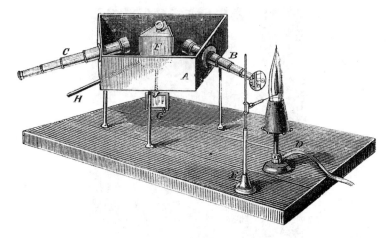

Figure 2-6. Spectroscope. The material to be tested is placed in the flame (D). Light from the flame then passes through a slit in the tube (B) to a prism (F), and is viewed through the telescope (C). From Henry E. Roscoe, *Spectrum Analysis* (London: Macmillan, 1869), 54.

There are so many ways that a sample can be compromised during the complex and delicate operations required to detect small quantities of a substance. Wary of mistaking trace impurities of a known substance for a new element, some chemists suggested that "radium" was actually barium contaminated by uranium. Or perhaps its previous exposure to uranium had made the barium radioactive, in the way that iron can be magnetized by contact with a magnet.

Some scientists even doubted that Becquerel's rays existed. This was not surprising, since the scientific literature overflowed with work of questionable quality on supposed new radiations. Giesel wrote to the Curies at the end of 1899 that "Unfortunately your beautiful discovery was hardly noticed at first, as people had become somewhat distrustful because of Le Bon's work.... even Röntgen did not believe in the existence of the Becquerel rays..."[5]

The Curies were delighted when the respected spectroscopist Eugène-Anatole Demarçay photographed radium chloride's spectrum (produced using a high voltage spark generator) and found a line which did not belong to any known element. This meant the line came from a new element, rather than from an impurity. But Marie Curie realized she needed to convince the skeptics on their own terms. The pattern produced by passing light through a prism might satisfy physicists, but many chemists still considered the spectroscope as a tool for physicists. Rather than using spectral patterns, chemists traditionally identified elements by their atomic weights. Marie Curie knew that if she could establish an atomic weight for radium, no reasonable scientist would deny its existence.

To obtain enough radium for weight measurements, the Curies would need large quantities of ore. Pitchblende was expensive, but after uranium was extracted from it, the remainders were discarded. The Austrian government owned a large pitchblende mine in the mountains near the village of St. Joachimsthal (meaning "St. Joachim's valley") in Bohemia (later Czechoslovakia). They were willing to send a generous supply of pitchblende waste to Paris if the recipients would pay for shipping. The Curies happily agreed. They asked Bémont to help them separate radium from the ore.

The storeroom was too small for such a large operation. The only facility available at the school was an abandoned shed with no floor. The roof leaked and the building provided little protection from heat and cold (Figure 2-7). These facilities shocked Wilhelm Ostwald, a famous German chemist, who once stopped by for a visit. He characterized the laboratory as "a cross between a stable and a potato-cellar, and, if I had not seen the table with the chemical apparatus, I would have thought it a practical joke."[6]

Figure 2-7. The shed where radium was discovered. Source: Musée Curie (coll. ACJC)/ Institut Curie.

The coveted shipment arrived early in 1899. In order to prevent toxic fumes from concentrating inside the shed, Marie performed most of the work outdoors in the building's courtyard. She processed approximately 220 pounds of ore in 45-pound batches, grinding, dissolving, boiling, stirring, filtering, pouring, and crystallizing, until the entire quantity had been separated into different chemical components. Each component was then tested for radioactivity.

The work would be exhausting even for a muscular man, but Marie's petite, pretty appearance belied her inner strength and determination. It would take nearly four years of heavy labor to produce enough radium to determine an atomic weight (one tenth of a gram, in 1902).[7]

That feat capped the Curies' work and crowned Marie's doctoral dissertation, which she presented in 1904. No longer eclipsed by the x rays, radioactivity was becoming a popular and exciting field in its own right. The Nobel Prize committee recognized this by awarding the 1903 physics prize jointly to Becquerel and the Curies for "'his discovery of spontaneous radioactivity'" and "'for their joint researches on the radiation phenomena discovered by Professor Henri Becquerel.'"[8]

Rutherford, Soddy, Particles, and Alchemy?

There was a spirit of adventure about him coupled with a dogged determination to reach his quest.

—Frederick Soddy, 1953

If it [atomic energy] could be tapped and controlled what an agent it would be in shaping the world's destiny!

—Frederick Soddy, 1904

For an alpha ray
Is a thing to pay
And a Nobel Prize,
One cannot despise,
And Rutherford
Has greatly scored,
As all the world now recognize.

A. A. Robb, c. 1919

RUTHERFORD AND THE RAYS

About the same time that Marie Curie was looking for radio-active elements, another physicist was investigating the rays. Ernest Rutherford, a bright and confident young man fresh from

New Zealand's Canterbury College, had left home with a scholarship to study with Joseph John (J. J.) Thomson, director of Cambridge's famed Cavendish Laboratory. Thomson, the English-speaking world's leading researcher on electricity, specialized in ionization (a process that creates electrified particles) and conduction in gases.

The son of a farmer and a teacher, Rutherford had became interested in physics and carried out researches as a student in New Zealand. His mother provided special encouragement, and he faithfully corresponded with her throughout his career. In Thomson's laboratory Rutherford continued his work on magnetism and high frequency electric waves (later known as radio waves).

Then Wilhelm Conrad Röntgen discovered x rays. The allure of these mysterious radiations changed Rutherford's career path forever. Caught up in the general excitement, Rutherford joined Thomson in investigating the electrical conduction which these rays produced in gases and became quite adept in ionization studies. Rutherford's work impressed his mentor and caught the attention of the wider scientific community. In 1898 he received an offer from McGill University in Montreal (Figure 3-1).

Though Rutherford would have preferred to remain in England at the center of British scientific activity, he was attracted by McGill's state-of-the-art physics laboratory. These fine facilities had been donated by Sir William MacDonald, an antismoker who, ironically, had made a fortune with tobacco. Rutherford (who happened to be a heavy smoker) accepted McGill's offer, and after he had settled in to his new surroundings, he began tests with Becquerel rays from uranium. He wanted to find out whether Becquerel rays were, as many suspected, a type of x radiation.

Figure 3-1. Ernest Rutherford. From the George Grantham Bain Collection, courtesy of the Library of Congress.

To trace the rays, Rutherford used the electrical methods per-fected at Cambridge, rather than the older photographic methods preferred by Henri Becquerel. This choice was crucial, for electri-cal measuring devices gave quantitative results, while photography (which involved waiting for an image to form on a photographic plate) was less precise and relied heavily on the observer's subjec-tive interpretation. First, Rutherford repeated Becquerel's experi-ments on refraction and polarization of rays from uranium. If these rays were a form of light, as Becquerel believed, it should be possible to refract and polarize them. But Rutherford could detect neither phenomenon.

Rutherford then tested the other possibility, that the Becquerel rays were a kind of x radiation. From his Cambridge work based on Thomson's electrical theory, Rutherford predicted the ionization that uranium's rays would create in air if they were a type of x ray. His measurements matched the predictions.

The rays were not uniform, as Becquerel and Gerhard Carl Schmidt had noticed earlier. Rutherford detected two components in them, one that could be stopped easily (which he called "alpha," after the first letter in the Greek alphabet, α) and another that could penetrate many layers of aluminum foil (which he called "beta," after the second letter in the Greek alphabet, β). This behavior was strikingly similar to a recent discovery by a French physicist. Georges Sagnac had shown in 1898 that x rays can generate new x rays when they strike matter. The secondary radiation, as Sagnac called it, was more readily absorbed than the original (primary) radiation. Rutherford believed that the Becquerel rays might be a mixture of primary and secondary x rays, with the betas being the primary radiation and the alphas the secondary radiation.

Early in 1899 Becquerel revisited his earlier experiments with uranium's rays. After he could not reproduce some of them and realized that he had misinterpreted others, Becquerel agreed that the rays he had discovered were more like x rays than light. So far as researchers in radioactivity were concerned, Becquerel rays were a form of x radiation.[1]

WHERE DID THE ENERGY COME FROM?

Radiation required energy to produce it. Primary x rays received their energy from the high voltage electric discharge tubes used to generate them. Secondary x rays received their energy from the primary x rays. What was the energy source for the Becquerel rays? Perhaps, suggested Marie Curie, space is full of some unknown radiation which supplies radioactivity's energy. When uranium and thorium absorb this radiation, they emit secondary rays as a type of phosphorescence. Curie's colleague Sagnac had recently found that x rays were absorbed best by the heaviest elements. Since thorium and uranium were heavy elements, Curie's idea fit well.

Thinking the radiation might come from the sun, the Curies compared uranium's activity at noon to its activity at midnight. At midnight, the radiation would need to pass through the earth to reach the uranium sample. If the earth absorbed some of this radiation, the midnight reading would be less than the reading at noon. However, the readings were identical. The Curies were so puzzled by the ongoing failures to find radioactivity's energy source that they were willing to consider a breach in the bedrock principle of energy conservation. Perhaps radioactivity was an exception to that rule. Sir William Crookes had also wondered whether radioactivity violated physical laws by gleaning its energy from motions of air molecules.

Two German physicists performed extensive tests in 1898–99 to locate radioactivity's source. Julius Elster and Hans Geitel were schoolteachers and respected physicists who had set up a private laboratory in Elster's home in the old medieval town of Wolfenbüttel. They had been friends since childhood and did most of their research together, so that the term "Elster-Geitel" signified a unit in the scientific community.

For years Elster and Geitel had investigated electrical effects in the atmosphere in order to understand weather phenomena, a popular research topic in the nineteenth century (Figure 3-2). When they learned of Becquerel's discovery, they wondered whether radioactivity might affect weather, and decided to

Figure 3-2. Julius Elster (left) and Hans Geitel (right). Courtesy of Archiv Fricke, Wolfenbüttel.

investigate the new phenomenon. Elster tried in vain to influence uranium's radiations with light and with heating. To see whether something in the air caused radioactivity, Elster and Geitel placed pitchblende and a uranium salt in a vacuum, then varied the air pressure. The radioactivity was not affected.[2]

To test the theory that an outside radiation caused radioactivity, Elster and Geitel took their radioactive samples to the bottom of a mine in the Harz Mountains south of Wolfenbüttel. If light or some other radiation caused radioactivity, surely more than a half-mile of earth would block some of it. Yet their sample went on sending out rays at the bottom of the mine just as it had at the surface.[3]

Becquerel rays seemed to behave like x rays. Since cathode rays produce x rays, could they also produce Becquerel rays? Elster and Geitel bombarded a piece of pitchblende with cathode rays, but its radiation did not change. Nor did sunlight have an effect. Over the years scientists tried to change radioactivity by applying x rays and radioactive radiations, to no avail.

If radioactivity's energy did not come from an outside source, the only alternative was an internal source. Many physicists suspected the atom was a system of charged particles. According to electrical theory, any disturbance to such a system would create further disturbances. Thomson suggested in 1898 that a heavy, presumably complex atom like uranium might rearrange its parts and cause radioactivity. A related idea was that the atom's parts were constantly moving, and that when they reached a particular unstable arrangement the atom would disintegrate.

Since they could not influence radioactivity, Elster and Geitel also surmised that some change inside the atom must cause it. This was an important conclusion, but it would be inaccurate to ascribe a modern interpretation to their words, written in 1899 before anything was known of atomic transmutation or nuclear power. Elster and Geitel envisioned something similar to an unstable chemical compound converting to a more stable form. Such a transformation would release energy and alter the atom's characteristics, as it would no longer be unstable. This did not mean that atoms of one element would change into another.

The same year Marie Curie posed a new and prescient possibility. Evolution was an exciting and controversial topic in the late nineteenth century. Perhaps, Curie suggested, radioactivity was a sign that the heavy elements were evolving into different forms. In just a few years two young researchers would confirm her hunch.

MATERIAL RAYS? DISCOVERY OF
THE BETA PARTICLE

Marie Curie was the first to suggest in print (January 1899) that Becquerel rays might be pieces of matter. Thomson had shown in 1897 that the cathode "rays" were actually small negatively charged particles, which he called "corpuscles." Scientists eventually preferred the term *electron*, first suggested in 1891 for the hypothetical unit of electricity. German physicists Emil Wiechert and Walter Kaufmann made similar findings and determined these particles were only about 1/2000 the mass of the smallest atom, hydrogen. If cathode rays were particles of matter, could the Becquerel rays also be material?

These scientists had used a common test to distinguish charged particles from nonmaterial radiation. A magnet can push, or deflect, moving charged particles from their paths, but has no effect on other radiations. The direction a particle moves when the magnet deflects it depends upon the particle's electric charge. Several researchers decided to apply a magnetic force, or field, to Becquerel rays.

Elster and Geitel had previously determined that a magnetic field could reduce certain electrical effects in air. They believed the magnet moved gas ions away from their measuring equipment, thus lowering the current recorded. Wondering whether ions produced by Becquerel rays would behave the same way, Elster and Geitel tried the experiment with uranium. Uranium's weak rays did not produce many ions, so they borrowed a sample of radium from Friedrich Giesel, who had just demonstrated radium at their local scientific society in Brunswick. Their magnet moved the ions that Giesel's radium produced.

But what if, in addition to moving the gas ions, the magnet deflected the Becquerel rays themselves? Elster and Geitel used a phosphorescent screen to detect the radium rays, which created a bright spot on the screen. They could not move the bright spot with their magnet. This seemed to show that Becquerel rays were not material particles, but rather more like x rays.

Giesel decided to check Elster and Geitel's work. Using a phosphorescent screen and photographic plates to record the paths of rays from radium and from a substance he believed to be polonium, Giesel found (in October 1899) that his magnet did deflect the rays. The direction of deflection changed when he changed the orientation of the magnet's poles. The rays must be material.

Figure 3-3. Stefan Meyer. Courtesy of the Austrian Central Library for Physics (Österreichische Zentralbibliothek für Physik).

Earlier in 1899 Giesel had demonstrated new radioactive materials at a scientific meeting in Munich. A young physicist from the University of Vienna was intrigued. The son of a writer and art collector, Stefan Meyer came from a distinguished Viennese middle-class family and grew up in a cultured and liberal-minded environment (Figure 3-3). Giesel agreed to loan samples to Meyer for his research on magnetic properties of the elements. Meyer also procured radium and polonium from the Curies.

Meyer teamed up with his colleague Egon Ritter von Schweidler, the son of an attorney. Schweidler had been investigating electricity in gases, which made radioactivity a natural fit for his next research (Figure 3-4). The two physicists decided to examine the electrical phenomena reported by Elster and Geitel. Unlike Elster and Geitel,

Figure 3-4. Egon Ritter von Schweidler. Courtesy of the Austrian Central Library for Physics (Österreichische Zentralbibliothek für Physik).

Meyer and Schweidler were able to deflect Becquerel rays. This result meant these rays were charged particles. From the direction of deflection they realized the particles were negatively charged.

Meanwhile in Paris, at the Museum of Natural History, the discoveries of radium and polonium had revived Becquerel's interest in radioactivity. Becquerel obtained samples from the Curies. After finding that radium's rays behaved like x rays insofar as reflection, refraction, and polarization were concerned, he compared the luminescence that Becquerel and x rays produced on phosphorescent minerals from the museum's collection. The results were ambiguous. Sometimes the effects were quite different, which could mean that the Becquerel rays were electromagnetic radiations of different wavelengths from the x rays. On the other hand, some minerals reacted to radium's rays very much as they had to cathode rays in earlier studies by his father, which suggested that Becquerel rays could be material particles.

To see whether x rays or cathode rays were the better match for Becquerel rays, Becquerel tested rays from both radium and polonium with an electromagnet. He then published the third paper in less than six weeks announcing that a magnet deflected the rays. The Becquerel rays were material, like cathode rays!

But not entirely. Only the more penetrating (beta) rays were deflected, according to Pierre Curie's report early in 1900. The rest of the Becquerel rays seemed to resemble x rays. The other researchers had not distinguished the alpha and beta rays in their experiments with magnets. These rays can be separated by inserting paper or cardboard in the rays' path, which will absorb the alphas but allow the betas to pass through.

Since electric forces also can deflect moving charged particles from their paths, Friedrich Ernst Dorn at Germany's University of Halle applied an electric force to radium's beta rays. The electric field, which would not affect the path of x rays, bent the path of the beta rays as expected.

These results pointed towards a new set of experiments. The distance a force field pushes a charged particle from its original path depends on both its charge and its mass. By sending charged particles through an arrangement which combines both kinds of force fields, the ratio between a particle's charge and mass can be determined. Becquerel succeeded first, getting values comparable to the known ratio of charge to mass for cathode ray particles. The cumulative evidence convinced most scientists that the beta rays were high speed cathode ray particles.

Rather than defeating the x-ray theory of radioactivity, the discovery of the beta particle strengthened it. When x rays strike matter, they produce secondary rays. Pierre Curie and Sagnac found that secondary rays carried a negative charge. Apparently, some of these rays were particles.

The parallels were unmistakable. Some secondary rays were negatively charged particles. The beta rays were also negatively charged particles, and identical to the cathode rays. It made sense to assume that Becquerel rays were a mixture of x rays and high speed cathode ray particles created by the x rays.

The beta particle's discovery gave physicists a tool for investigating an important theoretical question. Electromagnetic theory predicted that a particle's mass would increase with its velocity. The increase would be too small to detect with ordinary cathode rays. Perhaps the high speed beta particles would yield a measurable effect.

In 1902 Walter Kaufmann found that beta particles did increase in mass as predicted. Some took this as support for the radical idea that all mass was electrodynamic. Matter would be simply an illusion created by electricity in motion.

The discovery also encouraged radical speculations about radioactivity's cause. The Curies suggested a "ballistic hypothesis" for radioactivity in which the radium atom would gradually lose energy as it ejected negatively charged particles. "This viewpoint," they wrote in 1900, "would lead to the supposition of a mutable atom."[4] Their friend Jean Perrin, whose researches had helped verify Thomson's discovery of the electron, proposed a solar system type of model for the atom in which negatively charged particles rotated around positively charged "suns" of greater mass. Becquerel envisioned the atom as an aggregate of Thomson's corpuscles and larger positively charged particles. Yet, none of these theories was as revolutionary as the actual turn of events, for which we must go back to Rutherford in Canada.

THORIUM'S RAYS

After analyzing behaviors of uranium rays, Rutherford wondered whether rays from other radioactive substances would behave the same way. While uranium and thorium were commercially available, radium and polonium were difficult to procure, limiting choices for most researchers. This restriction was fortuitous for Rutherford, since thorium's strange behavior eventually led him to remarkable findings.

Rutherford's colleague at McGill, professor of electrical engineering R. B. Owens, examined rays from thorium oxide, as Rutherford had done for rays from uranium oxide. Owens noticed a radiation from thorium that penetrated thin sheets of aluminum foil. This radiation varied capriciously, but it seemed to depend upon air movements. Not sure what to make of this radiation but believing it was particulate, Rutherford called it an "emanation."

VANISHING RADIOACTIVITY

Meanwhile, a discovery in Germany threatened the assumption that radioactivity was permanent and unchanging. In August 1899 Giesel reported that polonium lost its activity over time.

Rutherford determined that thorium emanation also lost activity and set out to measure how fast its radioactivity decreased by measuring the current that thorium oxide created in air. Using an electrometer (similar to an electroscope but with a calibrated scale and a needle to record the current), he found that the current decreased exponentially with time. The activity fell to half its starting value in only one minute! This measure, the time it takes a substance to lose half its activity, became a standard in radioactivity. Later it was called the radioactive *half life* (Figure 3-5).

Thorium had another surprise in store for Rutherford: Everything the emanation touched became radioactive. This new radioactivity also weakened with time, but at a different rate from the emanation.

Rutherford first thought that the thorium emanation "excited" activity on other substances, in the way that light excites phosphorescence. He soon found that the activity behaved more like a layer of particles than something excited by an agent like light or magnetism. No matter what substance was used to collect the excited activity, it behaved consistently. It was attracted to the negative electrode, like a material substance, and could be removed by scrubbing or with strong acids. Careful measurements showed that the electrical effects created by the emanation and by the excited activity were proportional to each other. This suggested a causal relationship between them.

Rutherford failed to detect any emanation from a radium sample he received from Elster and Geitel. (The material probably was

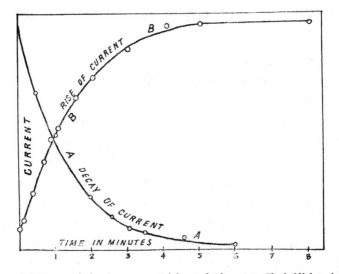

Figure 3-5. First graph showing exponential decay of radioactivity. The half life is about one minute. From Ernest Rutherford, "A Radioactive Substance Emitted from Thorium Compounds," *Philosophical Magazine* 49 (1900): 1–14, in *The Collected Papers of Lord Rutherford of Nelson* I, ed. James Chadwick (London: George Allen and Unwin, 1962), 224. Reproduced with permission of Ernest Rutherford's family.

mesothorium I, which behaves chemically like radium but does not produce an emanation.) However, in June 1900 Dorn announced that he had found a radium emanation. Rutherford ordered radium from Dorn's supplier, the chemical factory near Hanover owned by the Belgian Eugen de Haën, and began experiments with his student Harriet T. Brooks. The new sample produced an emanation. Assuming the emanation was a gas, Brooks and Rutherford measured its rate of diffusion into air.

During this time several chemists were getting puzzling results that turned out to be crucial for Rutherford's investigations. At first it seemed their problems with purifying uranium compounds had nothing to do with thorium's behavior. A well-known Hungarian chemist, Béla von Lengyel, precipitated a radioactive substance from uranium nitrate that acted like barium. The problem was that barium is not radioactive. What could this substance be? After getting similar results, Giesel noticed that his uranium nitrate lost some activity after the mysterious substance was separated from it. He had already observed (in 1899) that freshly prepared radium gains activity at first. What could be causing these irregularities?

In London, Sir William Crookes separated uranium nitrate into active and inactive components. Chemically, the active component did not behave like uranium. Crookes suggested that uranium's activity was due to an impurity, which he called "uranium X," "the unknown substance in uranium."[5] He suspected the impurity was radium.

In 1899 the Curies' colleague André-Louis Debierne, who had studied under the accomplished physical chemist Charles Friedel, had found another new active substance in pitchblende. He named it "actinium," probably from "actinic," the era's term for radiations that darkened a photographic plate. Debierne thought actinium resembled thorium chemically. He wondered whether thorium's radioactivity was really caused by traces of actinium. Becquerel removed an impurity from uranium which he thought might be actinium, but the uranium remained active. Apparently uranium's radioactivity did not come from actinium.

In 1901 Becquerel extracted what he thought was radioactive barium from uranium chloride. He repeated the extraction eighteen times, which caused the uranium to lose most of its radioactivity. It looked like Crookes could be right. Perhaps uranium's activity was caused by an impurity, most likely radium (which is chemically related to barium), and pure uranium was not radioactive at all.

Yet, Becquerel found that hard to believe. Although uranium ores from different places contained different kinds and amounts of impurities, these differences did not seem to matter. Uranium's activity did not depend on where it was mined. Then why did his chemical manipulations change uranium's activity, if an impurity was not responsible? Without resolving this puzzle, Becquerel set his uranium preparations aside and returned to his favorite subject, visible light.

Rutherford read these reports with interest. Could thorium's activity come from an impurity, as Debierne had suggested? To answer this question and to determine the nature of the emanation and the excited activity, Rutherford realized that he would need a chemist. He turned to McGill's young demonstrator in chemistry, Frederick Soddy. Fresh from Oxford's venerable Merton College, an institution with an illustrious past in the physical sciences, Soddy was bright, curious, and bold. He accepted the challenge (Figure 3-6).

Figure 3-6. Frederick Soddy. Courtesy of the Frederick Soddy Trust.

For their experiments Rutherford and Soddy used a highly sensitive electroscope that gave much more accurate results than photographic devices. First they searched for emanations from elements in thorium's group of the periodic table, since these elements share thorium's chemical properties. Only thorium gave off an emanation, which meant that other elements in thorium's group were not responsible. Apparently, thorium's emanating power was not connected to its chemical properties.

In order to uncover the emanation's chemical nature, Soddy tested it with various chemical reagents. When the emanation failed to react with any of the chemicals he tried, Soddy decided that it must be an inert gas. The gas came either from thorium itself or from the atmosphere. If the second possibility was correct, thorium somehow would have to make an inert gas in the air become radioactive.

TRANSMUTATION!

The first possibility raised a startling idea. Could thorium be transmuting itself into the emanation? Transmutation was an alarming notion, since it smacked of alchemy. Though long discredited by scientists, alchemy experienced a revival at the end of the nineteenth century. In France, four alchemical societies and a university of alchemy were founded. One of the societies published a monthly journal from 1896 to 1914. Several individuals reported converting one element into another, and in Glasgow a company was floated to change lead into gold, or mercury.

Most scientists disdained such efforts. When Soddy reportedly exclaimed, "Rutherford, this is transmutation: the thorium is disintegrating and transmuting itself into an argon gas," Rutherford replied, "For Mike's sake, Soddy, don't call it *transmutation*. They'll have our heads off as alchemists."[6] He did not want their work to suffer from guilt by association.

Before coming to such a radical conclusion, Soddy and Rutherford would have to find the emanation's source. Did it come from the atmosphere or from their thorium sample? They were not even sure that thorium itself was radioactive. Perhaps a minute impurity, analogous to Crookes' uranium X, caused thorium's radioactivity and its power to yield an emanation.

After many trials, Soddy concentrated something from thorium oxide that seemed to carry most of the radioactivity and also produced the emanation. As anticipated, the thorium left behind after the unknown substance was removed lost most of its radioactivity. He apparently had separated thorium X.

Soddy and Rutherford left for Christmas vacation, keeping their preparations in the laboratory. Meanwhile, Becquerel decided to reexamine his samples. Remembering Giesel's observation that freshly separated radium at first increases in radioactivity, Becquerel guessed he would find activity restored to his previously weakened samples. Borrowing a term used for electrical circuits and magnets, Becquerel envisioned the process as a type of "self-induction," where the uranium compounds somehow excited activity on themselves. His prediction was confirmed. The uranium had recovered its original activity, while the barium-like material had lost its powers. Becquerel published his findings in December 1901.

When Soddy and Rutherford returned to work, Becquerel's paper and a letter from Crookes about it had arrived in the mail. After reading these, Rutherford and Soddy eagerly checked their thorium preparations. Their results paralleled Becquerel's: Thorium X had lost nearly all of its activity, while thorium had regained its powers.

Not impressed by Becquerel's vague notion of self-induction, the McGill team set out to quantify the mysterious process by measuring how quickly thorium and thorium X lost and regained their activities. They called these measures the *rates of decay* and *recovery*. Soddy and Rutherford used an especially pure sample of thorium nitrate and an electrometer. It took about four days for thorium X to lose half its activity (its half life). During this time their thorium regained half of its original powers. Graphs of the data produced mirror-image curves, such that the sum of the activities of thorium and thorium X was always the same. These graphs could be represented as exponential functions (Figure 3-7).

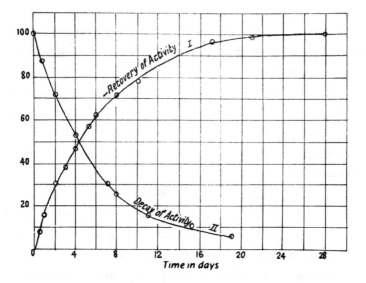

Figure 3-7. Decay and recovery of thorium and thorium X. From Ernest Rutherford and Frederick Soddy, "The Radioactivity of Thorium Compounds. II. The Cause and Nature of Radioactivity." *Transactions of the Chemical Society of London* 81 (1902): 837–860, on 841. In Ernest Rutherford, *In The Collected Papers of Lord Rutherford of Nelson* I, ed. James Chadwick (London: George Allen and Unwin, 1962), 439. Reproduced with permission of Ernest Rutherford's family.

While thorium X's activity decayed, thorium apparently was replacing it by producing more thorium X. Thorium X differed chemically from thorium, since it could be separated from it by a specific chemical reagent (ammonia). This meant that thorium was producing a substance chemically distinct from itself.[7]

Radioactivity, concluded Soddy and Rutherford soberly, was "a manifestation of subatomic chemical change."[8] The energy for radioactivity must come from a rearrangement within the atom. They avoided using the term "transmutation" with all its wild and disreputable associations, and asked the respected and well connected Crookes for his help in getting their controversial results published. These came out in 1902 in the prestigious *Transactions of the Chemical Society of London* (probably selected for Soddy's sake) under the neutral title "The Radioactivity of Thorium Compounds."

It might seem that Soddy's concern for his career would make him cautious. More likely, the reticence came from Rutherford, who may not have understood the strengths of Soddy's chemical demonstrations. Rutherford had dragged his feet on material interpretations of radioactivity. He had preferred the x-ray theory of Becquerel rays, which did not predicate any radical atomic change, and had been surprised by the discovery that the beta rays were material.

In contrast, Soddy had already speculated on transmutation before he began working with Rutherford. "The constitution of matter," he wrote in 1899, "is the province of chemistry, and little indeed can be known of this constitution until transmutation is accomplished." Soddy later claimed that he realized transmutation was happening when the thorium emanation turned out to be an inert gas, and that he had to convince Rutherford of this. In August 1903 he wrote to Rutherford that "Having failed utterly as I can see to make you realize the width of the gap between our recent work and anything preceding it I do not intend to attempt it in this letter ... "[9] It took further experiments with the Becquerel rays to resolve some technical details in Rutherford and Soddy's explanation of radioactivity and to make the theory's revolutionary nature clear to Rutherford.

A key issue was the sequence of changes which produced thorium X, the emanation, and the excited activity. Like Elster and Geitel, Rutherford supposed the radioactive process to begin with a rearrangement of the atom's internal parts caused by some unknown disturbance. The new arrangement was unstable, causing the atoms to radiate electromagnetic energy (like x rays) and perhaps beta particles (electrons). A second rearrangement would produce the unstable emanation, which would then send out rays and produce the excited activity. None of this supposed any radical change in the substances involved. Ray emission had become commonplace and was not connected with elemental changes in the substances that produced rays. However, Rutherford's sequence of events did not fit some details of his activity measurements.

A significant amount of rays emitted in radioactivity were alpha rays. Marie Curie had found (in January 1900) that these rays were absorbed more like projectiles than true radiations. Many physicists suspected that alpha rays carried a positive charge, including Hugh L. Callendar (Rutherford's predecessor at McGill), Robert J. Strutt (the future fourth Baron Rayleigh), Sir William Crookes, J. J. Thomson, Pierre Curie, and André Debierne. However, no one had been able to deflect a beam of alpha rays in a magnetic field, the crucial test for distinguishing charged particles from electromagnetic radiation.

Rutherford and Soddy failed to remove all of thorium's radioactivity when they separated thorium X from it. The residual activity consisted entirely of alpha rays. Knowing that some excited activities behaved as though they were positively charged particles, Rutherford decided to reinvestigate whether the alpha rays carried a positive charge.

The experiment would require a more powerful alpha ray source than the radium he had used earlier without success. Through Pierre Curie's intercession, Rutherford obtained a radium sample nineteen times more active than his previous source. He redesigned the measurement apparatus and borrowed a powerful magnet from McGill's electrical engineering department.

These improved resources brought success in the fall of 1902. Rutherford's apparatus deflected the alpha rays in the opposite direction from beta rays, proving they were positively charged particles. Their charge-to-mass ratio showed the alpha particles were much more massive than electrons, in fact comparable to the size of an atom. Their size (roughly 1,000 times the mass of the more readily deflected beta particle) had made them difficult to deflect. Rather than being a type of x ray, the alpha rays were positively charged particles.

The alpha particle's discovery brought the reality of atomic transformation home to Rutherford. It was easy to imagine that losing electrons or electromagnetic radiation would make little difference for an atom, but impossible to believe that ejecting atom-sized particles would not change it profoundly. Rather than being the consequence of some previous rearrangement inside the atom, ray emission was the cause of the subsequent changes. When Rutherford and Soddy revised their theory in 1903 to account for alpha emission, their problems with the sequence of changes disappeared. The atom-sized alpha particle was one of the disintegration products of the radioactive atom.

In 1902 McGill gained a liquid air plant, thanks to Sir William MacDonald's beneficence. With this low temperature apparatus

Rutherford and Soddy could condense both thorium and radium emanations (which liquified at slightly different temperatures), providing more evidence that these were material gases and that "radioactivity is accompanied by the continuous production of special types of active matter, which possess distinct and sharply defined chemical and physical properties."[10] In other words, transmutation was real.

Early in 1903 Soddy left McGill to work with the chemist Sir William Ramsay in London, while he looked for a permanent appointment.[11] Ramsay had recently added a new group to the periodic table, the inert gases. These included argon (which he codiscovered with the acclaimed physicist Lord Rayleigh), helium, neon, krypton, and xenon. Radioactive emanations, according to Soddy's work, should belong to that group. Both Ramsay and Soddy were eager to study the emanations, and leaving the provinces to work with the famous chemist was a good career move for Soddy.

A MISSED DISCOVERY

While Soddy and Rutherford were deciphering thorium's behavior and formulating a theory of atomic transmutation, French scientists studying radium's behavior were coming to different conclusions. The startling contrast in interpretations by the British and French teams resulted from their different views of the nature of science.

Shortly before Rutherford discovered thorium's excited radioactivity, the Curies found something similar with radium. At first they thought stray particles of radium were responsible. When experiments seemed to exclude this possibility, they concluded that radium was "inducing" radioactivity on other bodies. That term suggests something analogous to magnetic induction, where a magnet can create temporary magnetism in iron by contact. The term also is used for electromagnetic induction, where magnetic fields can "induce" electrical currents and electrical currents can "induce" magnetic fields, all via the universal ether. The induced activity persisted after the radium was removed. Ever ready to use his phosphorescence theory, Becquerel decided that the Curies' induced radioactivity must be similar to phosphorescence, with radium rays creating this activity in the same way that light creates phosphorescence.

Like Rutherford, the Curies found that the induced activity first increased, then decayed exponentially over time. Since experimenters had found radioactivity to be impervious to any force or circumstance they could devise, the Curies believed induced radioactivity was not genuine. They wondered whether other substances which appeared radioactive were only displaying induced activity.

One of these was polonium. When Giesel found that its activity decreased over time, it seemed the substance named for Marie Curie's beloved Poland might not be a new element at all. Puzzled and surprised, Giesel asked the Curies whether "by 'polonium' no ponderable matter is to be understood, but only an induced ether movement [electromagnetic radiation]?" Since polonium resembled bismuth chemically, Giesel decided that "the so-called polonium" was probably bismuth activated by a radium impurity.[12] Disappointed that polonium's radioactivity seemed spurious and believing that induced activity was only a minor effect of radioactivity, the Curies turned their attention to other aspects of the subject.

In 1901, after reading papers by Rutherford and Soddy on thorium's odd behavior, Pierre Curie decided to resume studies of radium emanation and excited activity. Like Rutherford, he paired with an excellent chemist, André Debierne.[13] Curie and Debierne found that radium induced more activity in a closed container than an open one and did not induce activity outside of a closed container. The induced activity was not confined to the path of the alpha and beta rays. It behaved suspiciously like a gas. The Curies had already noticed that pitchblende emitted a radioactive gas.

Yet, Curie and Debierne refused to surmise that the induced activity was a gas. This would be premature, they argued, since other explanations could be imagined. (They did not specify these other explanations.) Curie had made his position on proper scientific method clear in January 1902, after Becquerel had published a wildly speculative theory about emanation and excited radioactivity:

> In the study of unknown phenomena, one can make very general hypotheses and advance step by step with the assistance of experience. This methodological and sure progress is necessarily slow. One can, on the contrary, make bold hypotheses, where one specifies the mechanism of the phenomena; this manner of proceeding has the advantage of suggesting certain experiments and above all of facilitating reasoning by the use of an image. On the other hand, one can not hope to imagine thus *a priori* a complex theory in accord with experience. Precise hypotheses assuredly contain some error beside some truth; that ... forms only part of a more general proposition to which it will be necessary to return some day.[14]

This statement reflects a philosophical movement, positivism, that was popular on the continent during the nineteenth century. In science, this philosophy championed a noncommittal, mathematical approach and shunned visual models for phenomena. Two colleagues of Pierre Curie, Henri Poincaré and Pierre Duhem, were prominent spokesmen for positivism. A social variety of positivism which stressed human progress and the importance of education had captured Marie Curie's imagination in her youth.

Supporting their reluctance to adopt Rutherford's theory were experiments where the emanation did not behave like an ordinary gas. The emanation moved faster in capillary tubes than expected, appeared to be weightless, and did not produce a distinctive spectrum. Curie and Debierne suggested a thermal analogy for radioactivity, in which radioactivity's decay and recovery matched the behavior of a body reaching heat equilibrium. In correct positivistic form, they proposed neither a source for radioactivity's energy nor a mechanism for transmitting it. The emanation, according to Curie, was "the energy emitted by radioactive bodies in the special form in which it is stored in the gas and in the vacuum." The excited activity would be the radioactive energy under a different guise and with a different decay constant (a number used in the exponential function for radioactivity). Curie measured the radium emanation's activity over time and found that it decayed exponentially with a half life of about four days. "The current experiments show," he concluded, "that in the gas the energy is stored in a special form which dissipates itself according to an exponential law."[15]

Pierre Curie allied himself squarely with the nineteenth century's most universal and abstract physical theory, thermodynamics. Thermodynamics was a theory of heat regarded by many continental physicists as the epitome of a philosophically correct physical theory. Based on observation, general principles, and mathematical analysis, it supposed no concrete models or mechanisms to explain heat's behavior. Principles of thermodynamics could be extended to all energy forms, making energy a preferred referent for physical theories. For Curie, as for a significant minority of physicists, energy was more real than matter. "I see energy," he told his colleague and former student Paul Langevin.[16]

Pierre Curie set the tone for radioactivity research in Paris. Marie Curie, Becquerel, and Debierne did not hesitate to create models on their own. Marie Curie had proposed that radioactivity was caused by a profound change within the atom. Yet, these researchers deferred to Pierre Curie's views when they published jointly with him.

In contrast to the positivists, some nineteenth-century British physicists devised concrete, visual models to represent physical phenomena such as electricity. At that time Britain was experiencing major changes from its extended industrial revolution. The physicists' models were inspired by the industrial machinery around them and other objects that were easily visualized.

This approach was less popular on the continent. Positivism's influence and the greater value placed on presenting physics as a mathematical system made physical models less important to continental scientists, sometimes even undesirable. French and German scientists often hesitated to use a physical model without proof, while the British were willing to go forward with a hunch. The contrasting approaches of the Rutherford-Soddy team and Curie and Debierne starkly illustrated the difference.

This philosophical divide puzzled both camps. Early in the twentieth century, French physicist Pierre Duhem read Oliver Lodge's *Modern Theories of Electricity*. The author, a noted British physicist, used pulleys, pumps, weights, and gears as models for electrical phenomena. Duhem was bewildered. "We thought we were entering the tranquil and neatly ordered abode of reason, but we find ourselves in a factory," he lamented.[17]

After a visit from the distinguished French chemist Georges Urbain in 1914, the physicist Henry G. J. Moseley remarked that "the French point of view is essentially different from the English. Where we try to find models or analogies, they are quite content with laws." Rutherford observed that "continental people ... are quite content to explain everything on a certain assumption, and do not worry their heads about the real cause of the thing." Rutherford's debate with Curie and Debierne had forced him to confront this distinction. "I must, I think, say that the English point of view is much more physical and much to be preferred," he remarked.[18]

In spite of its successes with heat, thermodynamics was not a fruitful approach for radioactivity, which turned out to be an atomic phenomenon best suited to concrete physical models. Inhibited by his philosophy of science, Curie not only missed the discovery of atomic transmutation; he refused to accept the Rutherford-Soddy interpretation of radioactivity for several years.

Curie's own research forced him to change his mind. Even though he subjected radium to temperatures ranging from − 180°C to +500°C, its activity did not change. A chemical reaction could not be indifferent to such wide temperature variations. The next year Curie and his assistant Albert Laborde found that radium was several degrees warmer than its surroundings. Their measurements of the heat showed that a gram atom of radium would generate an amazing 22,500 small calories per hour! Such a large amount of energy, they conceded, might come from subatomic transformation. Curie and another assistant, Jacques Danne, confirmed that radium emanation diffused and condensed like an ordinary gas.

The *tour de force* came in August 1903. Until then radium's scarcity had severely restricted research on radium emanation. That year Giesel's firm began marketing highly purified radium bromide (50% by weight) at an affordable price. Stunned to find radium for sale in a London shop for only eight shillings per milligram, Soddy promptly purchased twenty milligrams to use in Ramsay's laboratory for experiments on radium emanation.

Their first attempts to isolate the emanation failed dismally, but spectral tests showed a helium line in the mixture of gases that radium produced. Rutherford happened to be in London at the time. Soddy took him to the shop that was selling inexpensive radium, where Rutherford purchased about thirty milligrams, then loaned it to Soddy and Ramsay. With this larger sample, Soddy and Ramsay obtained nearly all of helium's visible spectrum.

Next, Soddy and Ramsay combined the gases produced by both radium samples. They did not detect any helium, but after waiting several days, helium's spectrum appeared. Apparently, radium emanation was producing helium!

Because of concerns about contamination, Soddy and Ramsay modified their apparatus and took extra steps to purify the gases from the radium samples, which they sealed in a glass tube. In a few days, helium's spectrum appeared in the tube, confirming helium's presence beyond doubt.

Helium was a chemical element distinct from radium. It could not have come from anywhere but the sealed tube. This experiment vividly revealed atomic transmutation in action. "Helium could be, according to these results, one of the disintegration products of radium," conceded Curie.[19]

After Ramsay and Soddy overcame the difficulties of isolating radium emanation, Ramsay enlisted his colleague John Norman Collie to identify its spectrum. A versatile chemist, Collie had worked with Rayleigh on inert gases and published on helium's spectrum. Collie found lines in the emanation's spectrum that did not match any lines from the known elements. This meant that radium emanation was a new element, produced from radium by atomic transmutation. It resembled the inert gases newly identified by Ramsay. Later it was named "radon."

When he visited London in June 1903 to lecture at the Royal Institution, Curie brought a radium sample to the British chemist and low temperature expert James Dewar. Dewar liquified a gas produced by the radium, which turned out to be helium. The emanation and helium were clearly material, making Curie's hypothesis of special energy forms superfluous. Curie dropped his philosophical objections to the transmutation theory and incorporated the Rutherford-Soddy theory of atomic transmutation into his 1904–05 class at the Sorbonne.[20]

REACTIONS

Rutherford and Soddy's theory created quite a stir. Publications on radioactivity more than doubled in 1903, and even the popular media noticed. "We are on the crest of a wave of interest," remarked Soddy. Clemens Winkler, discoverer of the element germanium (named to honor his homeland, Germany), wrote of "the excitement about radium, which now pervades the world," and the British physicist Joseph Larmor predicted that Rutherford "may be the lion of the season for the newspapers have become radioactive."[21]

Certainly the enthusiasm spread through the Western-educated world. The field was so successful that two specialized journals were founded in 1904, *Le Radium* (Radium) by Pierre Curie's former student Jacques Danne and *Jahrbuch der*

Radioactivität und Elektronik (Yearbook of Radioactivity and Electronics) by German physicist Johannes Stark. In the same year Soddy accepted a position as independent lecturer in physical chemistry at the University of Glasgow, with the understanding that he would create a research school. The Chemical Society of London asked Soddy to write a radioactivity section for their *Annual Reports on the Progress of Chemistry*, which he contributed from 1904 to 1920. The Italian physicist Augusto Righi wrote a textbook of radioactivity based on the Rutherford-Soddy theory, and the Japanese physicist Hantaro Nagaoka proposed an atomic model which featured disintegration.

Physicists overwhelmingly accepted the Rutherford-Soddy theory, but most chemists were not enthusiastic. They distrusted a theory based on evidence from the physicists' electroscopes and spectroscopes. The transmutation hypothesis depended on experiments with quantities of matter much too small for chemical tests. After Marie Curie and others found an atomic weight for radium, chemists reluctantly accepted it into their world. But the idea of minuscule quantities of decay products that disappeared like ghosts before the experimenters' eyes was too much to swallow. As late as 1914 Marie Curie remarked on the skepticism caused by "the disconcerting character of the new chemistry... of the Invisible which seems to derive from phantasmagoria."[22]

In spite of the Chemical Society of London's progressive role, many chemists continued to ignore radioactivity because they viewed it as a subfield of physics. They would not pay much attention to the transmutation theory, since the chemist "is not given to elaborate theories and is usually adverse to speculation," and they did not think the theory would affect their everyday work. To the average chemist, radioactivity and the Rutherford-Soddy theory were irrelevant. "It is sad to think," reflected Soddy, "that on this, the greatest discovery ever made in their science, the chemists had nothing of any value to say, and agreement went by default of any possible objection to it."[23]

This indifference, and the cases where individuals proposed alternatives to the transmutation theory, are not unusual responses to a new theory. What is remarkable is that the transmutation theory was accepted so quickly. In spite of Soddy's concerns, the opposition was weak and ineffectual. The German chemist Willy Marckwald observed in 1908 that "its rapid successes are perhaps without precedent in the history of science."[24]

The main reason the transmutation theory was readily adopted was that scientists had been primed by mounting evidence that atoms were not simple and indivisible. Atoms could break down into charged objects called *ions*. They contained tiny electrically charged particles, or electrons, which appeared as cathode rays and beta rays. Electrons were involved in atmospheric electricity and the electrical effects produced by light (the photoelectric effect). They caused the Zeeman effect and were believed to create atomic spectra.

Growing knowledge of ions and electrons merged with scientific speculations about the evolution of the chemical elements and long-standing suspicions that these elements were made of smaller building blocks. All these signs that atoms were complex made it plausible that, by a change in structure, one type of atom could transform into another.

The period's widespread fanciful ideas about alchemy, energy, and disintegration of matter were but faint distorted shadows of the cutting edge of experimental physics and wonders yet to appear. Though these ideas eased acceptance of transmutation theory by the general public, they did not inspire the scientific theory.

ATOMIC ENERGY?

The first measurements of radium's heat astounded researchers, and their reactions spilled over into the wider community. Rutherford and others had estimated radioactivity's heating effects from alpha ray energies, but obtained much lower values than Curie and Laborde. 22,500 calories *per hour* from a gram atom of anything seemed fantastic, but the researchers reporting this value had impeccable credentials. Soon, several scientists confirmed the results.

This amount of heat was comparable to the energy produced by burning one gram atom of hydrogen—and unlike radium, which seemed to generate heat endlessly, the hydrogen would be consumed in the process. To burn hydrogen was to produce the most powerful chemical reaction known. Yet, without stove or fire, radium could boil a little more than its weight of ice-cold water in an hour!

Rutherford and Soddy calculated the total energy that might be released when a gram of radium completed all of its transformations. Their estimate of 100,000,000 to 10,000,000,000 calories made atomic energy seem a likely source for the sun's power.

Emission of fast-moving particles and light from atoms had already convinced scientists that the atom was a storehouse of energy. Radioactivity's unending output had pressed the question, How much energy? The results from France as well as Rutherford and Soddy's estimates raised the ante by several orders of magnitude. The energy stores were much larger than anticipated and seemingly limitless. This revelation invited speculations. Could this energy reservoir be tapped by humanity? Would atomic energy be used for good or harm?

"Moonshine," was Rutherford's characteristic response to speculations about harnessing the atom's energy. He pictured this energy as coming from movements of the particles constrained within the atom. All attempts to tap it had failed. J. J. Thomson's laboratory had tried in vain to break up atoms by bombarding them with the powerful x rays and cathode rays that their excellent equipment supplied. To the end of his life Rutherford believed the problem of tapping this energy was too difficult to solve in the foreseeable future. In 1933 he stated that

> we cannot control atomic energy to an extent which would be of any value commercially, and I believe we are not likely ever to be able to do so. A lot of nonsense has been talked about transmutation. Our interest in the matter is purely scientific, and the experiments which are being carried out will help us to a better understanding of the structure of matter.[25]

Pierre Curie judged differently, suspecting the problem would eventually be solved. With some apprehension, he noted that, in the hands of criminals, radium could be quite dangerous. But he preferred to be optimistic: "I am one of those who believe ... that mankind will derive more good than harm from the new discoveries."[26]

Soddy became excited about possibilities for using the atom's energy to help humanity. He envisioned that this energy could "transform a desert continent, thaw the frozen poles, and make the whole world a smiling Garden of Eden."[27] Later he wrote extensively about energy and its role in society. Unique among radioactivity's pioneers, Soddy immediately connected radium's heat with conversion of mass into energy, in line with the electromagnetic theory of matter. If some of radium's mass changed into energy, comparing the weight of a radium sample with the weight of its decay products should give the amount of mass that vanished during transmutation and reappeared as energy.

Albert Einstein published a theory of mass-energy equivalence in 1905. Einstein arrived at his famous result (later expressed as $E = mc^2$) by starting with assumptions that differed from those used to calculate a relationship from electromagnetic theory. Though Einstein later was credited with the idea that mass could be converted into energy, the idea itself was not new and Einstein's theory did not receive much attention at first. Neither form of the mass-energy hypothesis proved possible to test before the 1930s.

Though by 1906 most scientists believed radioactivity's energy came from inside the atom, it was still possible that radioactive atoms garnered some energy from an outside source. The mysterious force of gravity had been proposed in 1902 by Adolf Heydweiller in Germany. His countryman Robert Geigel published experimental tests of gravity's effect on radioactivity in 1903, which most physicists found unconvincing. Still, the idea was intriguing. The British physicist Sir Arthur Schuster linked radioactivity to an eighteenth-century theory of gravitation by the Swiss physicist, mathematician, and cleric Georges Louis Lesage.

In 1906 the Curies' colleague Georges Sagnac revisited the idea that gravity might power radioactivity. Could radioactive substances, the heaviest elements known, have a special knack for absorbing gravitational energy? Sagnac tested Lesage's hypothesis but could not get meaningful results. Other physicists obtained negative results. The hypothesis of atomic energy emerged victorious in the debates on radioactivity's source. The consequences for humanity would not become clear for several decades.

TRAGEDY

On April 19, 1906, Pierre Curie left home for a normal workday. He would never return. He was killed a few hours later when he absent-mindedly stepped into a Paris street in the path of a horse-drawn vehicle. Curie's death shocked the scientific world and the broader public, especially in France where he was a national hero. Marie Curie was devastated. She found it impossible to tell her children of the death and would not speak of it even after they became adults.

Perhaps feeling remiss at having neglected the Curies' research needs, the Faculty of Sciences at the University of Paris took the unprecedented step of appointing Marie Curie to Pierre Curie's post. Marie carried her chemical experience to the laboratory where, in addition to the physical aspects of radioactivity, she promoted study of its chemical complexities.

As Marie Curie tried to cope with her grief, radioactivity became even more of an obsession for her. This work seemed to accompany every thought and act. She judged endeavors on the basis of their benefit to radioactivity and, by extension, to science. Curie's natural attention to detail and desire for completeness now directed her focus to the intricacies of chemical separations, to the composition of rays, and generally to tying up loose ends in the field. The important but uncreative task of creating international standards for radioactivity measurements occupied much of her energy. The imaginative, enthusiastic girl with wide interests transformed into a withdrawn, depressed, and chronically ill woman who focused on narrow issues and avoided groundbreaking researches.

Yet, Curie's focus can be viewed not simply as a psychological reaction to her loss, but as reflecting a broader view of radioactivity which, in addition to physics and chemistry, encompassed industry, medicine, and metrology. Curie wanted her laboratory's efforts to be useful not only for research, but also for radium therapy and the radium industry. The laboratory's extensive work on measurement and standardization was essential for these applied fields. Her collaborations with industry and medicine benefited all parties concerned.

Despite her prolonged grieving, Marie Curie's sorrow did not weaken her intellectual acumen, her social conscience, and her desire to help students pursue scientific investigations. A former student reflected that "it was not difficult to notice the flame which was burning within her... the flame of idealism and enthusiasm."[28] She directed a research laboratory with many workers, some of whom made significant contributions to radioactivity and carried their knowledge to other nations. Much later, her student Marguerite Perey discovered a new element. Curie's daughter Irène, who later won the Nobel Prize, was the most notable fruit of Curie's formal and informal mentoring.

During 1920–31 Curie published work on theoretically important issues like the Compton effect, the radioactive decay constant, and relations between alpha and gamma ray spectra. Throughout her career she retained her curiosity about radioactivity's cause, her interest in new developments, and her knack for insightful speculation.

In 1911 Marie Curie became the first person to receive two Nobel Prizes. The Swedish Royal Academy of Sciences awarded the chemistry prize to Curie "in recognition of her services to the advancement of chemistry by the discovery of the elements radium and polonium, by the isolation of radium and the study of the nature and compounds of this remarkable element."[29]

Some concluded that Curie had received two prizes for the same work. Perhaps an element of sympathy influenced the decision, as she was seriously ill in 1911 and the subject of a public scandal (because of her close friendship with physicist Paul Langevin). However, Curie received the first prize before she had accomplished the remarkable feat of isolating radium. She had advanced polonium researches considerably since 1904. The importance of radioactivity's medical applications was also a legitimate consideration for the 1911 award.

MORE RAYS

In 1900 Paul Villard, a French physicist who had also trained as a chemist, was experimenting with radium's penetrating (beta) rays. Since beta rays were thought to be beams of negatively charged particles, he expected them to behave like cathode rays. Surprisingly, Villard found a difference. Radium's penetrating rays made faint marks on photographic plates in a position inaccessible to cathode rays.

Suspecting these traces were created by rays different from the beta rays, yet more penetrating than the alphas, Villard decided to apply the definitive test for charged particles. He sent radium rays through a magnetic field.

The magnet broke the beam into two parts. One part was deflected like beta rays. The other part continued in a straight line, oblivious to the magnet.[30] This second portion, evidently uncharged, traveled through nearly ten inches of air, an aluminum plate, and several pieces of paper before it finally left its mark on a photographic plate. Rays from a more powerful sample loaned by the Curies could penetrate the radium's glass container, paper, and three millimeters of lead. Alpha rays could not have passed through all these materials. Beta rays would have been deflected by the magnet.

If these powerful rays were neither alpha nor beta rays, what could they be? X rays were a likely candidate, and Villard concluded as much. To continue the naming pattern established by Rutherford, these rays were called gamma rays, since gamma (γ) is the letter that follows alpha and beta in the Greek alphabet. The gammas were a highly energetic type of x ray.

The gamma rays were joined by the delta rays in 1904, named after the fourth letter of the Greek alphabet, δ. The deltas were slow-moving electrons that appeared during alpha ray emission, and were included in the family of radiations for about ten years. They turned out to be electrons that alpha rays knocked out of materials in their surroundings, a type of secondary radiation, rather than true spontaneous radiations that came from the radioactive substances themselves.

THE ALPHA PARTICLE

Unlike beta rays, alpha rays could be blocked by only a sheet of paper or a short distance of air. Researchers first assumed these less penetrating rays were also less important, a mere secondary radiation provoked by the more powerful beta rays. This theory collapsed after Soddy inferred and McGill student A. G. Grier demonstrated that uranium and thorium emitted only alpha rays. Beta rays did not appear until further down the disintegration chain, so they could not have excited the alpha rays. Instead, the alpha rays initiated the whole decay sequence. Rutherford accepted this scheme after he found the alpha rays were heavy charged particles.

If the alpha particle was as large as an atom, what element might it represent? Helium was a likely candidate, since radioactive minerals often contained helium. When Soddy and Ramsay showed that radium gave off helium as it disintegrated, Rutherford guessed that the helium was an accumulation of alpha particles. To prove this he would need to find the alpha particle's mass and compare that with helium's atomic weight. Deflection experiments could provide the ratio between the alpha particle's charge and its mass. If Rutherford could find the charge of an alpha particle, he could use the ratio to calculate its mass.

Rutherford found the total charge carried by radium's alpha particles per second. To find the charge for a single particle, he would need to divide the total charge by the number of alpha particles which the radium sample emitted per second. Counting these alpha particles proved a daunting task.

One possible method would be to count the flashes of light, or scintillations, which alpha particles created when they struck zinc sulfide screens. The problem was that no one could be sure that every particle registered on the screen, and that each particle triggered a single flash.

Rutherford decided to use the electrical methods which had served him so well. He would count alpha particles indirectly by measuring the ionization they produced in a container. Unfortunately, the ionization was weak and the measurements erratic.

In 1907 Rutherford left McGill to take a position at the University of Manchester. Sir Arthur Schuster, a prominent physicist with an independent income, had resigned his position there in order to bring Rutherford back to the center of British scientific activity. After arriving in Manchester Rutherford recruited Hans Geiger, a German researcher in the laboratory, to develop a device that would magnify alpha particle ionization to a level where it could be measured consistently and accurately.

The alpha particles ionized air by colliding with atoms and molecules with enough momentum to knock electrons out of them. The released electrons would create an electric current, which

Rutherford had been trying to measure. To magnify this current, Rutherford and Geiger used a principle discovered at Cambridge by Rutherford's friend John S. Townsend. Townsend realized that an electrical voltage properly applied would accelerate the electrons released by ionization so that they would collide with other atoms and dislodge more electrons. Eventually a cascade of electrons would reach the measuring device and cause its needle to move, so that the experimenter could record the current.

Rutherford and Geiger started with a design for an instrument later known as an ionization chamber, developed by Townsend's associate P. J. Kirkby. The chamber was a brass cylinder with a window to admit alpha particles. A wire passing through the cylinder's center would collect the electrons dislodged when an alpha particle passed through the chamber and conduct them to a recording device. The chamber was designed so that each alpha particle would produce a separate burst of electrons, making it possible to count individual alpha particles by counting the number of times the measuring device recorded an electron burst.

The idea was simple, but the details were challenging. Geiger had to adjust the gas pressure and the voltage in the counter and arrange the apparatus so that each alpha particle ionized exactly one molecule. The ejected electrons would need to move through the apparatus so they could be recorded, and the electrometer needle had to reset itself quickly after each surge of electricity to be ready for the next. If sparks developed, the experiment would be ruined. Since sparks ionize air and allow electricity to be transferred more readily than before sparking, any measurements would be worthless. Sparks would also interfere with resetting the electrometer between measurements.

After many trials, Geiger succeeded. He was able to determine the number of alpha particles the radium sample sent out per second. Geiger and Rutherford then computed the charge each alpha

particle carried and the particle's mass. These results matched Rutherford's hypothesis that the alpha particles were helium atoms carrying a charge twice that of the beta particle but opposite in sign.

The match was not proof, for it might be that something else carried that mass and charge. To determine the chemical nature of the alpha particles, Rutherford needed to find their spectrum. Rutherford enlisted Thomas Royds, a research fellow in his laboratory, to collect alpha particles from radium emanation held inside a specially designed glass tube. Otto Baumbach, Rutherford's skilled glassblower, was able to make tubes thin enough for alpha particles to pass through the glass, yet strong enough to withstand the atmosphere's pressure. Tests of the alpha particles collected in another tube revealed helium's spectrum. The alpha particles were positively charged atoms of helium.

Rutherford assumed that alpha particles were ready-made components of radioactive atoms. They flew out of atoms when some sort of atomic instability caused an internal explosion. Since the alpha particles were helium, helium must be a building block of radioactive elements. Perhaps, Rutherford speculated, helium was a building block of other elements as well.

In 1908 Rutherford received the Nobel Prize for Chemistry "'for his investigations into the disintegration of the elements, and the chemistry of radioactive substances.'" "I am very startled," he wrote to the German chemist Otto Hahn, "at my metamorphosis into a chemist." Going along with the spirit of the award, he titled his Nobel Prize lecture "The Chemical Nature of the α-Particles from Radioactive Substances." Of all the transformations he had encountered, the quickest, he quipped to the audience gathered in Stockholm for the prize ceremony, "was his own transformation in one moment from a physicist to a chemist."[31]

The Radioactive Earth

I said Lord Kelvin had limited the age of the earth, *provided no new source of heat was discovered.*

—*E. Rutherford*

Results of these observations appear to be best explained by the assumption that a radiation of very high penetrating power enters our atmosphere from above …

—*Victor F. Hess, 1912*

THE PROSPECTORS

Discovery of a new element was a sure route to fame. The announcements by the Curies and by Gerhard Carl Schmidt sent hopeful imitators scurrying around the globe searching for radioactivity. Perhaps polonium and radium were only the first of new elements awaiting discovery. Even if they were unique, it might be profitable to find more sources of radioactive materials. The needed equipment was simplicity itself—just an ordinary leaf electroscope would do.

Since Benjamin Franklin's time scientists had known that air conducted electricity slightly. But why? This question had preoccupied Julius Elster and Hans Geitel for decades. Meteorology was a popular research topic in the nineteenth century, and the air's conductivity was a major puzzle. Elster and Geitel collected data on the electrical state of air from various locations at different times of day and seasons of the year, hoping to find patterns that would lead them to the source of atmospheric electricity and a better understanding of weather. They developed a portable electroscope to use in their researches.

New theories of electrical conduction proposed in the late nineteenth century explained how the atmosphere became electrified, but not why. Electricity was carried by particles (ions), which were created when radiation broke molecules apart in the process called ionization. Could Becquerel rays be the agent that ionized air, causing the atmosphere's electricity?

Elster and Geitel set out to answer this question by matching changes in the air's ionization with the characteristic ionization patterns produced by different radioactive elements. Soon they were followed by dozens of researchers, who combed the earth, air, and seas for radioactive materials. Radioactivity turned up everywhere—in springs, wells, rocks, mud, the ocean, rain, snow, even volcanos. Instead of new elements, the prospectors found only thorium and radium, plus the radioactive gases these elements produced. Scientists were surprised to find these rare materials so widely dispersed. No wonder the air was charged with ions. Elster and Geitel decided that radioactivity caused the bulk of the atmosphere's ionization, and most other scientists concurred.

If radioactive materials in the earth created the atmosphere's electricity, the electrification should be greatest at ground level and decrease with height. In 1910 Father Theodor Wulf brought an electroscope to the Eiffel Tower in Paris to compare measurements at the tower's base and top. The ionization was less at the top of the tower than at ground level but higher than expected.

Could something besides radioactive materials be creating ions in the air?

Balloons could carry scientists higher than any tower. Several researchers took balloon flights to measure the air's ionization, including Albert Gockel from Switzerland, Karl Kurz and Karl Bergwitz from Germany, and Victor Hess from Austria. During an ascent the ionization would first decrease as expected, but then it would level off as the balloon rose higher, sometimes even increasing. Gockel and Hess suspected that some unknown radiation source was causing these strange results.

Hess made seven balloon flights in 1912 to track the anomaly. His electroscope's readings fell at first as the balloon rose, but after 1,000 to 2,000 meters of ascent (0.6 to 1.2 miles) they began to increase, slowly at first, then dramatically as the balloon climbed as high as 4,500 to 5,200 meters (2.7 to 3.1 miles). Other researchers confirmed these results.

Apparently radioactive materials did cause most of the air's ionization close to the ground, but at high altitudes something else became important. A mysterious ionizing radiation called "high altitude radiation" by Hess and later named "cosmic rays" joined the physicists' roster of invisible rays inviting investigation.[1] As with radioactivity, the study of cosmic rays eventually merged with the fields of nuclear and particle physics.

HOW OLD IS THE EARTH?

Radioactivity transformed the long-standing debate on the earth's age. Early views based on a literal reading of the Bible assumed the earth was no more than a few thousand years old. During the eighteenth and nineteenth centuries findings in paleontology and geology extended these estimates to tens and even hundreds of millions of years. Charles Darwin and others claimed that the evolution of living creatures also required an extended time span.

But if the earth were so old, why had it not cooled to a very low temperature in the interim? Most nineteenth-century scientists believed the earth started out as a hot body. Much of the heat would have radiated away if the earth were more than a few thousand years old. Yet, volcanos revealed an immense store of heat within the earth. The theory of thermodynamics, believed to contain physics' most fundamental laws, could not account for this heat so dramatically demonstrated.

The British physicist William Thomson, later honored with the title "Lord Kelvin," had tackled these problems during the previous century. Assuming that the earth was originally hot, he used heat theory to compute its rate of cooling and (from that) its age. Kelvin's first estimate in the 1860s of 100,000,000 years was difficult to reconcile with geology; his 1897 estimate of about 24,000,000 years was impossible.

As for the earth's future, prospects were dismal. Unless some unknown source were found to replenish its heat, the earth would eventually become too cold to support life. Worse, the entire universe would gradually run down and lose its available energy, disappearing into oblivion with the so-called heat death of the universe. Only an outside energy source could extend the earth's age back into time and save the universe's future. This seemed a futile hope.

In just a few years this bleak picture was transformed. Radioactivity promised to be Kelvin's unknown energy source, supplying the missing heat for an ancient earth. Pierre Curie and Albert Laborde showed that radium gave off prodigious amounts of heat, with no end to its powers in sight. Since the earth's crust was riddled with radium and other radioactive substances, all bets were off on the earth's age and its future prospects.

To get answers for the earth's age, researchers turned to radioactive rocks. Knowing from laboratory measurements how quickly a radioactive element decayed, they could predict how much would be left in a rock and how much of its decay product would be created after any time interval. After measuring the amounts of the element and its decay product contained in the rock, scientists could estimate how long the transformation process had been going on, which would be the rock's age. This technique, *radioactive dating*, was later used to estimate the age of fossils and other artifacts.

Helium's presence in radioactive ores, and experiments showing that radium produced it, made helium an early choice for radioactive dating attempts. In 1905 the British physicist Robert J. Strutt set the age of a radium salt at a mind-boggling 2,000,000,000 years, while Rutherford estimated in 1906 that ores he tested were at least 400,000,000 years old. Helium's propensity to escape from rocks over time limited this method's accuracy.

In 1907 Bertram B. Boltwood suggested that lead was the final product of uranium disintegration. Several scientists analyzed rocks to compare the amounts of uranium and lead they contained. These figures would allow them to estimate ages. Unlike the gas helium, lead was more likely to remain in rocks indefinitely, making these estimates more reliable. One specimen yielded the amazing age of 1,600,000,000 years! Over the next decade, researchers continued to unravel the decay chains for the radioactive elements, making it possible to compare amounts of these elements and their decay products in rocks.

Early experimenters found that radioactivity can color glass, gems, and mineral crystals. The responsible agent turned out to be the alpha particle. Many minerals contain small amounts of radioactive elements which eject alpha particles and create colored regions along their paths. Since the rays go out in all directions, the colored regions are spherical. The radius of the sphere in a mineral equals the range of the alpha particles that created it. Cross sections of these minerals show circular "halos."

Researchers tried to use halos to find the ages of minerals, but did not get accurate results. However, measurements revealed something very important. Alpha particle ranges determined from ancient mineral halos were the same as contemporary values found by other methods. Researchers already knew that radioactivity did not change during the few years they had been studying it. Since an alpha particle's range is related to the decay rate (and half life) of the substance that emits it, the mineral halo measurements showed that radioactivity had not varied over eons of time. So far as anyone could tell, radioactive decay constants and half lives were fixed properties of radioactive matter.

A NEW PROPERTY OF MATTER?

Up to this time radioactivity had been found only in the heaviest elements, but no one knew why. Some scientists wondered whether radioactivity was a property of matter, like magnetism. If so, every element should be radioactive, just as every element reacts to a magnetic field. Perhaps atoms of all elements decayed, but only a few elements emitted radiation strong enough to be detected. The supposed discovery of "rayless" changes in 1903, changes inferred from decay products but not accompanied by rays, encouraged these speculations.

First suggested by Sir Arthur Schuster and by J. J. Thomson, the notion that radioactivity was a universal property of matter fit well with current electrical theory and ideas about the atom, which was commonly believed to contain electrically charged particles in motion. According to James Clerk Maxwell's electromagnetic theory, moving charged particles will send out radiation whenever their speed or direction of movement changes. Some physicists interpreted this to mean that all atoms should be slightly radioactive. Gustave Le Bon popularized a vague and imaginative version of this idea based on his belief that all matter was disintegrating.

Universal radioactivity did not lose its appeal when the alpha rays turned out to be particles instead of radiation. After he found that alpha particles were not detected if they traveled below a minimum velocity, Ernest Rutherford suggested that rayless changes actually involved slow-moving alpha particles. Unseen by us, the world of "ordinary" matter could be slowly decaying by alpha emission.

In 1906 physicists Norman Campbell and Albert B. Wood claimed that potassium and rubidium, two metals very different from uranium, radium, and thorium, were weakly radioactive. Tests eliminated possibilities that the effects came from trace impurities (such as radium or uranium) or from outside sources like light. The radioactivity of potassium and rubidium seemed genuine, but no one found evidence to justify extending the hypothesis of universal radioactivity to the rest of the periodic table.

Traces of radioactive elements were so widely dispersed and the effects sought were so weak that it seemed impossible to prove that all elements were radioactive. Most scientists set the possibility aside, since there was no way to completely exclude impurities or to record an undetectable alpha radiation. Though attractive, the idea of universal radioactivity was not justified by the evidence, and by 1913 it was not widely supported. Whatever else radioactivity might be, it probably was not a universal property of matter.

Speculations

It would appear as if the rate of transformation of the atoms depends purely on the laws of probability ...

—*Ernest Rutherford*, 1913

... the essence of which [the law of probability] is that you know nothing about it!

—*F. Soddy*, 1953

EARLY THEORIES

By the early twentieth century, scientists had pinned down radioactivity's mysterious source to the atom, yet the atom's interior was itself a mystery. Many assumed it contained the same negatively charged particles found in beta rays, cathode rays, secondary radiation, the Zeeman effect, and the photoelectric effect. Since atoms ordinarily had no charge, scientists believed they must include some source of positive charge which balanced the negative charge.

Many researchers thought the negatively charged particles were moving rapidly while they were inside the atom. This assumption could explain how beta particles acquired their energy. Several scientists designed atomic models which contained moving charged particles. From Maxwell's theory it followed that charged particles inside the atom should radiate energy, as radioactive elements were known to do.

Using Maxwell's theory of radiation to explain radioactivity led to a problem. If atoms were constantly losing energy by radiation, they should eventually collapse or explode. Yet most atoms are stable. A way out of the dilemma was to suppose atoms would not disintegrate until their parts reached some unstable configuration. Julius Elster and Hans Geitel first suggested that an unstable rearrangement of an atom's inner parts might cause radioactivity, though they were thinking only of a superficial process like ionization.

In Britain, both Sir Oliver Lodge and J. J. Thomson suggested that the radiating particles (Thomson's corpuscles) might create a disturbance inside the atom that made it unstable. Thomson supposed that an atom would become unstable after a corpuscle's velocity had dropped below a critical value, just as a spinning top becomes unstable as it slows down. The unstable atom might explode and eject particles or even divide itself into two or more parts.

Ernest Rutherford became this theory's champion, basing radioactivity on a gradual change inside the atom caused by electrons draining energy. If the theory were correct, older atoms would be more likely to disintegrate than newly formed atoms since the older ones had spent more time losing energy. But nothing, including age, seemed to affect radioactivity. Rutherford himself had codiscovered the exponential law of decay, which meant that the odds for an atom to disintegrate had nothing to do with its previous history.

Frustrated with this quandary, Rutherford admitted in 1912 that "it is difficult to offer any explanation of the causes operating which lead to the ultimate disintegration of the atom." Perhaps, he suggested lamely, atoms of the same substance began life with different degrees of stability.[1]

RADIOACTIVITY AND PROBABILITY

Since radioactivity resisted all efforts to affect it, many suspected it was a random process, governed by the laws of chance. Mathematical analysis supported these hunches. The exponential functions used by Frederick Soddy and Rutherford for radioactive decay were used to describe probabilistic (chance) behavior. Soddy explicitly noted the connection. In Germany, physicist Emil Bose reasoned in 1904 that the radioactive decay constant measured the probability for an atom to decay.

This did not mean scientists believed radioactivity operated outside of physical laws. They treated probability theory as a kind of descriptive agnosticism, something that described a phenomenon without explaining it. Nearly all scientists assumed radioactivity followed basic principles of physics, but realized their incomplete knowledge limited them to predicting average behaviors of large numbers of atoms, rather than behaviors of individual atoms.

Graphs of radioactive decay strikingly resembled a well-known probabilistic relationship for energies of molecules in a gas, the Maxwell-Boltzmann distribution law. This law was named after Maxwell and the Austrian mathematical physicist Ludwig Boltzmann (Figure 5-1). Before he developed his electromagnetic theory, Maxwell had analyzed the motion of gas molecules. The results became known as the *kinetic theory of gases*, after a Greek word meaning "to move."

According to kinetic theory, movements of gas molecules can be described mathematically as random events. Nineteenth-

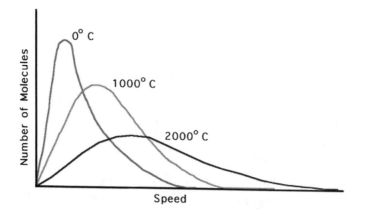

Figure 5-1. Maxwell-Boltzmann distribution curves for three temperatures. Copyright © 1998, Washington University in St. Louis. Reprinted with permission.

century scientists generally assumed the individual particles followed known physical laws, but since the particles were too small to be observed, only their average behavior could be predicted. Boltzmann extended Maxwell's theoretical work on kinetic theory.

Boltzmann was a brilliant and passionate theoretical physicist. He championed the idea that atoms were real, in opposition to many continental physicists, who viewed the atom as merely a convenient image that had not been proven to exist. Boltzmann developed a strong tradition of physics in Vienna based on his own and Maxwell's theoretical work. His colleague Franz Exner, an experimental physicist and director of the Vienna Physical Institute, was philosophically inclined to apply those theoretical principles not only to physics, but also to the humanities, economics, and other aspects of culture.

Boltzmann and Exner were loved and respected leaders in their fields. Their work on probability and randomness planted seeds of doubt in a physics that assumed a completely predictable universe.

They mentored a generation of physicists at the University of Vienna, including future prize winners and other prestigious leaders in the field.

At the turn of the century Vienna was an exciting intellectual and cultural center where philosophy, physics, and politics were discussed in coffee houses. Sigmund Freud developed his psychoanalytical theories there, and artists like Gustav Klimt and Egon Schiele experimented with new ways to portray the city's emotional ferment. A mixing place for different peoples and traditions, Vienna was a congenial venue for supporters of socialism, feminism, and other progressive ideas.

Notwithstanding his successes and the stimulating intellectual environment, Boltzmann was tormented by spells of depression. He increasingly felt alone and professionally embattled. Tragically, in 1906 Boltzmann took his own life. Not long afterwards, experiments on Brownian motion (random movement of microscopic particles) and radioactivity vindicated his belief in the atom's reality. Boltzmann's students and colleagues carried on his scientific legacy. One was Egon von Schweidler, who earlier had codiscovered the magnetic deflection of beta rays. In 1905 he turned the Viennese mathematical heritage towards radioactivity theory.

If radioactivity was a random process, individual atoms would not decay in a predictable sequence. Instead, they would disintegrate at widely varying time intervals. Schweidler developed a formula for predicting variations in these time intervals, called "fluctuations." Experimenters could find the fluctuations by recording the times that rays from radioactive samples entered their measuring devices. Over the next few years researchers in Canada, Austria, Germany, Sweden, and England observed "Schweidler's fluctuations," confirming that radioactivity was a random process that followed laws of probability.

In 1910 Harry Bateman at Cambridge applied von Schweidler's theory to alpha particles. Rutherford, Hans Geiger, Ernest Marsden, T. Barratt, and Marie Curie tested Bateman's work by recording flashes that alpha particles created when they struck a phosphorescent screen. The random appearance of these scintillations matched the theory. Phosphorescent screens (made first with zinc sulfide, which registers alpha particles, and later with barium platinum cyanide for beta particles) became one of radioactivity's most basic tools.

In retrospect, Schweidler's analysis was seen as a turning point for physics, since randomness eventually became a core principle in modern theories. But at the time, confirmation of von Schweidler's theory only deepened the mystery. Scientists could have defied common sense and decided that laws of cause and effect did not operate inside the radioactive atom. Quite reasonably, most preferred the view that causes existed but were not observable. They yearned to find an explanation for the exploding atoms. Soddy remarked that the randomness made it "difficult to construct any model of the disintegrating mechanism."[2] How could anyone construct a model for radioactivity if there was no way to predict when a particular atom would disintegrate?

KINETIC MODELS OF THE ATOM

One way to construct a model under these circumstances was to be vague about details. Based on the average behavior of large numbers of moving particles rather than on their individual behaviors, kinetic theory fit the bill. In 1904 Soddy opened the door for kinetic models of radioactivity by suggesting that an atom's interior must have parts which behaved like the molecules of a gas. He supposed that "A chance collocation of the parts" in this atom would lead to atomic decay.[3] Several scientists used this idea to develop theories of radioactivity based on heat and the kinetic theory of gases. In these models the decay process was assumed to follow physical laws, but was contingent on a random process that preceded it.

In kinetic theory, temperature is a measure of energy carried by gas molecules. In 1912 Harold A. Wilson, a physicist at McGill who had worked in Thomson's laboratory at Cambridge, proposed a parallel concept for the atom's interior, an atomic temperature. Instead of representing the energy carried by gas molecules, atomic temperature would measure the energy of alpha particles within the atom. An atom's decay rate would depend upon its atomic temperature: The higher the energy, the greater the odds that an atom would decay. The German physicist Eilhard Wiedemann had developed similar ideas for luminescence earlier. The idea of atomic temperature fit well with the era's near worship of thermodynamics, and leading spokesman Henri Poincaré found it convincing.

Wilson tried to unravel a puzzling mathematical relationship formulated by Geiger and John M. Nuttall in Rutherford's Manchester laboratory in 1911. Geiger and Nuttall showed that the faster a substance decayed, the farther its alpha particles would travel. To represent this relationship they developed an equation known as the "Geiger-Nuttall law." This law intrigued researchers, who suspected it held a key to the mysterious decay process. Marie Curie considered it "the first that has been found for the kinetics of the atom;" André Debierne called it "a first relation of the intraatomic kinetics and thermodynamics."[4] Their intuitions proved correct much later after a new theory called *wave mechanics* was used to explain the rule.

In 1912 Debierne suggested that the atom contained unspecified particles which behaved like gas molecules. Their random movements would create myriads of situations inside the atom, some of which might cause an explosion. Located deep within the atom's core, these particles would be shielded from external forces.

Debierne compared the atom to a planet with an atmosphere and a core. The atmosphere would produce phenomena which could be affected by outside forces, like chemical and electrical actions. The core would produce radioactivity and would not be influenced by outside forces. It would remain hidden, except for its violent, volcano-like eruptions. In rough outline, Debierne's speculations sketch the picture of the atom which took shape over the coming decades.

Rather than a gas of randomly moving particles, the Berlin theoretical physicist Friedrich Lindemann envisioned a core, or nucleus, of rotating particles. When the particles reached some unspecified critical configuration, the atom would decay. Unlike previous theorists, Lindemann included the *quantum* (a quantity of energy that depended upon frequency) in his hypothetical nucleus.

After examining Debierne's theory, Marie Curie proposed a variation of the chance configuration theme which included a remarkable conceptual leap. Perhaps, she suggested, the radioactive atom was like a box containing a small opening through which a particle can escape. With a large number of boxes, the exit (decay) will follow the law of chance, even though the process behind it is simple.[5] Curie's speculation foreshadowed the wave mechanical interpretation of alpha particle decay proposed in 1928. According to this theory, alpha particles moving inside a radioactive atom can occasionally escape when they reach the atom's surface, even if they do not carry much energy. The process is random, as it would be if the atom were a box with a hole in it.

Though imaginative and suggestive, there was no way to test these early models. Radioactivity's cause was as mysterious as ever. Meanwhile, chemists had unraveled key principles from the tangle of radioelements and their decay sequences, opening another route to understanding the atom.

Radioactivity and Chemistry

It appears that chemistry has to consider cases, in direct opposition to the principle of the Periodic Law, of complete chemical identity between elements presumably of different atomic weights.

—*Frederick Soddy*, 1911

THE RISE OF RADIOCHEMISTRY

Radioactivity was first considered a branch of physics, since it involved the study of rays. Established disciplinary confines could not contain this rapidly evolving field, which required both physics and chemistry in its pursuit. The researcher took on the role of physicist or chemist, depending upon the task at hand.

The difficult work of separating new radioactive bodies and determining their chemical properties required the chemist's expertise. A chemist, Gustave Bémont, collaborated in radium's discovery. The Curies drew the chemist André Debierne into their research, yet Debierne later received a doctorate in physics. Marie Curie, trained in physics, chose to perform many of the necessary chemical separations herself. This work led her to a lifelong interest in the chemical behavior of radioactive substances.

Ernest Rutherford quickly recognized chemistry's value. Rather than involving himself closely with the details, he relied on the chemist Frederick Soddy. In addition to his excellent professional skills, Soddy's deep understanding of chemistry enabled him to recognize atomic transmutation when he confronted it.

In spite of chemistry's importance for the new field, most chemists shunned this line of research. They were unfamiliar and uncomfortable with the physicists' methods. Chemists identified elements by their atomic weights, yet it was impossible to find atomic weights for minute traces of materials that could vanish before anyone could weigh them. Some doubted whether researchers were finding new elements at all. Most likely, ran this view, they were finding small quantities of known elements and misinterpreting their results. The physicists' electrical techniques seemed to be creating "a chemistry of phantoms."[1] If radioactivity included chemistry, it would have to be a different sort than what most chemists practiced.

That new chemistry became known as "radiochemistry," a field that crossed traditional boundaries. At first, its novelty could be a hindrance for an aspiring radiochemist who wished to rise up the academic ladder. Logically, the unconventional specialty fell under the realm of physical chemistry. Soddy's scientific reputation and connections made him an ideal candidate for the University of Glasgow's new position in this area. Yale University instituted the first chair specifically for radiochemistry in 1910, naming Bertram B. Boltwood as the first occupant.

The radiochemists were an independent breed, pursuing a topic ignored by most of their colleagues. Many came from German speaking regions like Berlin, Munich, and Vienna, since Germany led the world in chemical research, education, and industry. Time and success were on their side. By the 1920s radiochemistry was a respected field with many practitioners.

RADIOACTIVE GENEALOGY

As researchers identified more and more products of radioactive decay, these became known collectively as "radioelements." Scientists struggled to make sense of the growing number of radioelements by organizing them into families and finding their positions in the radioactive family trees, known as "decay series." Researchers adopted a system for identifying the different generations, which made it easier to notice similarities between different decay series. The name of the parent element (radium, uranium, thorium, or actinium) would be followed by a word or a letter of the alphabet that showed where each radioelement belonged in its series. In 1904 Rutherford presented this scheme:[2]

Radium → Radium emanation → Radium A → Radium B
→ Radium C → Radium D → &c.
Thorium → ThX → Thorium emanation → Thorium A
→ Thorium B → Thorium C → Final product
Uranium → UrX → Final product
Actinium → Actinium X(?) → Actinium emanation
→ Actinium A → Actinium B → Actinium C (final product)

The sequences became more complex as researchers added new decay products to these series, but the general naming pattern was maintained.

Though Rutherford placed uranium and radium in different families, he and Soddy suspected that radium descended from uranium because only uranium minerals contained it. In 1904 Boltwood measured the amounts of uranium and radium in various uranium minerals. Each mineral contained a constant proportion of radium to uranium, which bolstered the idea that these elements belonged to the same family. Soddy and Boltwood both showed that radium did not come directly from uranium, but instead from some unknown uranium decay product. A few years later Boltwood found the missing element, which he named "ionium." Berlin chemist Willy Marckwald (who previously had found a new element that turned out to be polonium) and his student Bruno Keetman made the same discovery independently. Now the uranium and radium decay sequences could be combined, so that

Uranium → Uranium X → Ionium → Radium
→ [all the radium descendants]

CHEMISTRY OF THE IMPONDERABLE

Confronted with minute amounts of matter which often decayed too rapidly to test by chemical analysis or with a spectroscope, researchers responded by developing new techniques. One technique for "the chemistry of the imponderable"[3] was electrolysis, where the experimenter sent an electrical current through a solution using two wires attached to a voltage source. The ends of the wires were attached to plates or rods used as electrodes. If the electrodes were placed in a metal salt solution, the metal would deposit on an electrode, a process known as electroplating. With the right choice of materials, a metal will deposit on a plate or rod placed in the solution even if no outside voltage is applied.

Different metals will deposit on an electrode at different rates, depending upon which metal is used for the electrode. Results of such electrochemical experiments had given chemists a tool for analyzing unknown substances that could work for radioactive materials. By observing how readily a radioactive decay product would deposit onto different metals, chemists could find out which element the radioactive substance most closely resembled. Pioneers of electrochemical methods included Ernst Dorn, Marckwald, Friedrich Giesel, the Hungarian-born chemist Friedrich von Lerch, and George B. Pegram, a physicist at Columbia University.

Another method to disentangle a mixture of radioactive elements required separating the parent of the product most difficult to separate. As the parent element decayed, it would produce the desired element. Experimenters could also separate radioelements by heating the mixture, since the components would evaporate and condense at different temperatures.

One separation method used a principle of physics called the *conservation of momentum*. When an atom sends out an alpha or a beta particle, the atom will recoil, or kick back, with an equal and opposite momentum. J. J. Thomson predicted this process in 1901, but it was not observed until 1904, when Rutherford's student Harriet Brooks experimented with a wire exposed to radium emanation (radon). For many years, no one exploited recoil's potential as an analytical tool.

Otto Hahn rediscovered recoil in 1909 while experimenting with actinium. Rutherford's colleagues Sidney Russ and Walter Makower at the University of Manchester showed that both alpha and beta decay produced recoil. Researchers used recoil to separate radioactive substances in a decay chain, which led to discoveries of new decay products.

Scientists characterized radioelements by their radiations and decay rates, which were physical properties. To determine a radioelement's chemical properties, researchers used several techniques other than electrolysis. For standard chemical analysis, a solution of the radioelement was mixed with a solution of a known element (sometimes called the "carrier"). Then the researcher added a chemical that would react with the known material to form an insoluble solid, or precipitate.

The experimenter tested the precipitate and the solution for radioactivity. If the radioelement's activity appeared in the precipitate, it had reacted like the known element and must resemble it chemically. If the radioelement remained in the solution, it had not reacted and thus did not resemble the known substance. The experimenter could then try to precipitate it with a different chemical.

Once the radioelement to be tested had been precipitated, chemists would use other chemicals to separate it from the carrier. By repeating these processes, and sometimes using other methods as well, they could purify the unknown substance and do further tests with it.

A variation of this method used crystallization to determine whether a radioelement was related chemically to a known element. The experimenter put both substances in solution together and waited for the solution to form crystals. If the radioelement resembled the known element, it would become part of the crystal structure.

INSEPARABLE RADIOELEMENTS

Chemists soon ran into problems with their chemical separation methods. Some substances refused to be dislodged from their carrier elements. As these cases multiplied, radiochemistry became increasingly confusing.

First there was radiolead. Around 1900 several observers noticed that lead taken from uranium minerals was radioactive. Karl Andreas Hofmann and Eduard Strauss in Munich believed they had found a new radioelement, while Giesel thought lead's activity was induced by traces of radium. Since lead was not radioactive when extracted from minerals devoid of uranium or radium, several scientists dismissed the idea of a new element in lead.

The controversy over radiolead drew in many experimenters, including André Debierne; the Hungarian chemist Béla Szilard; Marie Curie's student H. Herchfinkel in Paris; Stefan Meyer, Egon von Schweidler, and V. Wolfl in Vienna; Julius Elster and Hans Geitel in Germany; Norman R. Campbell and Albert B. Wood in Cambridge; and Steward J. Lloyd in Alabama. Tests showed the active substance in lead was neither uranium nor radium. It seemed to be a radium decay product, perhaps radium D; but no one could separate it from lead.

In 1912 Rutherford told his new student, a chemist from Hungary, "If you are worth your salt, you separate radium D from all that nuisance of lead."[4] György (Georg) von Hevesy attacked the problem with youthful confidence; but after almost two years of fruitless attempts, he had to admit failure.

Radiothorium was another problem. While searching the famous hot springs of Baden-Baden in southern Germany in 1904, Elster and Geitel found a new radioactive substance that gave off thorium emanation. The next year Hahn made a comparable discovery while he was in London doing a research stint in Sir William Ramsay's laboratory. Curiously, he could not separate the unknown substance (which he named "radiothorium") from thorium. An Italian, Gian A. Blanc, reported similar failures. Rutherford and his friend Boltwood were skeptical about Hahn's new radioelement. Boltwood dismissed radiothorium as "a new compound of Th-X and stupidity."[5]

Rutherford and Boltwood relented after Hahn joined Rutherford's laboratory at McGill and convinced Rutherford that radiothorium was not a figment of his imagination. Just the same, radiothorium resisted all efforts to separate it from thorium.

The element ionium, which Boltwood had discovered in uranium ore, also refused to be separated from thorium. Knowledgeable chemists like Boltwood, Hahn, Marckwald, and Keetman tried without success. Even the prominent Austrian chemist Carl Auer von Welsbach, an expert on the hard-to-separate rare earth elements (and inventor of several popular lighting methods), failed to separate ionium from thorium. To complicate matters, Keetman could not separate ionium from actinium X.

ISOTOPES

The inseparable elements were a major problem for the periodic table of the elements, a classification scheme which chemists had used since the nineteenth century. During that century several chemists had devised ways to arrange the elements in a diagram or chart, but the Russian chemist Dmitri Mendeleyev developed the version eventually adopted. In Mendeleyev's table, atomic weight determined chemical properties. When he ordered the elements by atomic weight, they fell into chemical families, for instance, the alkali metals and the halogen gases. But if two or more elements were so much alike that no one could separate them, where should they be placed in the periodic table?

Worse, the table was running out of space for new radioelements. A few positions were still empty, and any elements heavier than uranium could be inserted at the end of the table. But the new radioelements could not be heavier than their parents, and there were too many of them to fit into the table's remaining spaces. Perhaps, suggested Keetman in 1909, several elements could share one position in the table. Such elements should be nearly equal in atomic weight since their chemical properties were so similar. The idea was not completely original, as Sir William Crookes had suggested something comparable back in 1886 to explain rare earth spectra, but scientists had devised another way to place the rare earth elements in the table[6] (Figure 6-1).

Unlike the inactive elements in the periodic table, the radioelements had ancestries. All seemed to originate from three parents: uranium, thorium, and actinium. As the parents decayed, they produced lines of descendants, some with close resemblances. For instance, all three decay series included inert gases, the "emanations" later named radon, thoron, and actinon. Observers noticed parallels between uranium X, thorium X, and actinium X, as well as between radiothorium and radioactinium. These analogous substances usually appeared at the same position in each series and often sent out the same kinds of rays (Appendix 2).[7]

In 1909 Swedish chemists Daniel Strömholm and Theodor (or "The," pronounced *Tay*) Svedberg did extensive chemical tests to find out where the radioelements belonged in the periodic table. They were struck by the analogies between the three decay series. Could it be that these analogous substances, which no one had been able to separate, did share positions in the periodic table? If so, Mendeleyev's scheme would need revision. Each "element" in his table would be a mixture of several elements of nearly identical atomic weight.

There was no way to find the atomic weights of the troublesome radioelements directly, since they were produced in such minute

PERIODIC TABLE OF THE ELEMENTS

	GROUP O.	GROUP I.	GROUP II.	GROUP III.	GROUP IV.	GROUP V.	GROUP VI.	GROUP VII.	GROUP VIII.
Hydrogen 1.008									
	Helium He 3.99	Lithium Li 6.94	Beryllium Be 9.1	Boron B 11.0	Carbon C 12.00	Nitrogen N 14.01	Oxygen O 16.00	Fluorine F 19.0	
	Neon Ne 20.2	Sodium Na 23.00	Magnesium Mg 24.32	Aluminium Al 27.1	Silicon Si 28.3	Phosphorus P 31.04	Sulphur S 32.07	Chlorine Cl 35.46	
A	Argon A 39.88	Potassium K 39.10	Calcium Ca 40.07	Scandium Sc 44.1	Titanium Ti 48.1	Vanadium V 51.0	Chromium Cr 52.0	Manganese Mn 54.93	Iron Fe 55.84 Cobalt Co 58.97 Nickel Ni 58.68
B		Copper Cu 63.57	Zinc Zn 65.37	Gallium Ga 69.9	Germanium Ge 72.5	Arsenic As 74.96	Selenium Se 79.2	Bromine Br 79.92	
A	Krypton Kr 82.92	Rubidium Rb 85.45	Strontium Sr 87.63	Yttrium Yt 89.0	Zirconium Zr 90.6	Niobium Nb 93.5	Molybdenum Mo 96.0	—	Ruthenium Ru 101.7 Rhodium Rh 102.9 Palladium Pd 106.7
B		Silver Ag 107.88	Cadmium Cd 112.40	Indium In 114.8	Tin Sn 119.0	Antimony Sb 120.2	Tellurium Te 127.5	Iodine I 126.92	
A	Xenon Xe 130.2	Caesium Cs 132.81	Barium Ba 137.37	[Lanthanum La 139.0	Cerium Ce 140.03	Praseodymium Pr 140.6	Neodymium Nd 144.3		Samarium Sa 150.4
	Europium Eu 152.0	Gadolinium Gd 157.3	Terbium Tb 159.2	Dysprosium Dy 162.5	Erbium Er 167.7				
	Thulium Tm 168.5	Ytterbium Yb 172.0	Lutecium Lu 174.0]						
B		Gold Au 197.2	Mercury Hg 200.6	Thallium Tl 204.0	Lead Pb 207.10	Bismuth Bi 208.0	(Polonium)	—	Tantalum Ta 181.5 Tungsten W 184.0 Osmium Os 190.9 Iridium Ir 193.1 Platinum Pt 195.2
A	Radium Emanation 222.	—	Radium Ra 226.0	Actinium	Thorium Th 232.4	Uranium X_2 (Brevium)	Uranium U 238.5		

Only the four spaces marked ⸺ are vacant places.

Figure 6-1. Periodic table, 1914. The elements are ordered by atomic weight rather than by atomic number. Four places in the table are empty. From Frederick Soddy, *The Chemistry of the Radio-Elements. Part II.* (London: Longmans, Green and Co., 1914), 10.

quantities. Perhaps sensitive spectral measurements could distinguish them. In 1912 Soddy's student Alexander S. Russell, then working in Rutherford's Manchester laboratory with spectroscopist R. Rossi, searched unsuccessfully for new spectral lines from a thorium—ionium mixture. Eduard Haschek and Franz Exner, director of Vienna's Institute for Radium Research, also observed only thorium's lines. Likewise, they found nothing new in radiolead's spectrum.

Investigations of a radium decay product named "mesothorium," discovered by Hahn in 1907, triggered the puzzle's resolution. Mesothorium resembled radium so closely that manufacturers used it as a radium substitute. In order to protect the interests of the Berlin chemical firm that supplied him with thorium preparations, Hahn kept the process for producing mesothorium secret. Soddy decided to work out the process himself.

In the meantime a chemical manufacturer asked Marckwald to find the amount of radium in a "radium" preparation. The sample turned out to be mostly mesothorium. After unsuccessfully trying to separate out the mesothorium, Marckwald concluded in 1910 that mesothorium was chemically "completely similar" to radium.[8]

Soddy also failed to separate mesothorium from radium, but went a step further than Marckwald. Soddy had reviewed the literature on radioactivity for the Chemical Society of London's *Annual Reports* since 1904 and was thoroughly familiar with the research on inseparable elements. When he encountered mesothorium's inseparability, he was ready to take a radical step. In 1911 he proposed that the inseparable elements were not only similar, but identical.

For nearly a century, the idea that each chemical element had a unique atomic weight had been chemical dogma. Soddy, who had already overthrown the belief in unchangeable elements with the transmutation theory, was now willing to discard the primacy of atomic weight. His solution to the problem of inseparable substances fit so well that scientists accepted it quickly.

By 1911 the groups of identical substances included thorium X, actinium X, radium, and mesothorium; thorium, radiothorium, ionium, and uranium X; the emanations from radium, thorium, and actinium; and radium D and lead. Soddy reasoned that chemically identical substances must share the same place in the periodic table. A family friend translated "same place" into Greek as *iso topos* (literally "equal place") for Soddy, who converted this to "isotopes." *Isotopy* became the principle that a single chemical element can exist in more than one form, or isotope. Using this concept, the new radioelements could be accommodated in the periodic table. Scientists could then use the table to predict properties of missing members in the radioactive decay series.

This episode illustrates how problematic it can be to credit one person with a discovery. Marckwald's "completely similar" essentially meant "identical." His collaborator Keetman had suggested that inseparable elements might share a single position in the periodic table. If we replace "inseparable elements" by "isotopes," Strömholm and Svedberg's 1909 conclusion that the chemical elements were actually mixtures of inseparable elements with different atomic weights is identical to Soddy's position.

Yet, credit for the discovery fell to Soddy. Soddy was well placed and better known than the others and had an exceptional record in the field. He received the 1921 Nobel Prize for Chemistry for discovering isotopy and for his other radiochemical researches. Svedberg did not pursue the topic of isotopes. He was rewarded with the 1926 prize for his work on colloid chemistry.

Hevesy turned his failure to separate radium D from lead into an ingenious method for investigating processes in plants, animals, and humans. He showed that a small quantity of a radioactive isotope, called an "indicator" or "tracer," could be used to mark its inseparable non-radioactive element. Researchers could find out how the element traveled and where it concentrated by tracking the tracer's radioactivity. Later, the radioactive tracer method gave valuable information to physiologists and medical researchers, as well as agricultural scientists, chemists, metallurgists, and industrial scientists. Hevesy's work was rewarded by the 1943 Nobel Prize for Chemistry.

DISPLACEMENT LAWS

While isotopy was emerging from the immense volume of radiochemical research, chemists were noticing relationships between the rays that radioactive substances emitted and chemical properties of the products. Most radioelements could not be isolated for standard chemical tests, but electrolysis would reveal their electrochemical properties. In 1906 both von Lerch in Vienna and Richard Lucas in Leipzig developed a generalization about the radioelements. Each determined that successive products in a decay series become increasingly electronegative. This so-called law of Lucas or von Lerch guided researchers until exceptions were found in 1912. Radiochemists then searched for a more accurate way to express the relationship between a radioelement's place in a decay series and its position in the periodic table.

The main action centered around the laboratories of Soddy in Glasgow and Rutherford in Manchester, with influence from Marckwald in Berlin. Four researchers published a version of what became known as the radioactive displacement laws: Russell and Soddy separately in Glasgow, Hevesy in Manchester, and the chemist Kasimir Fajans in Karlsruhe, Germany. Their professional paths intertwined, and some shared mentors. Russell was introduced to radiochemistry by Marckwald. Fajans, Russell, and Hevesy studied radioactivity at Manchester under Rutherford; and Russell also worked with Soddy, who in turn had worked with Rutherford. Research by Soddy's demonstrator Alexander Fleck was instrumental for the solution.

Fajans' route to the discovery was especially colorful. A native of Poland, Fajans directed a research group at Karlsruhe's Technical University, where he endured good-natured teasing about his clumsiness in the laboratory. This ineptness did not interfere with his chemical astuteness. For a long time he had been puzzling over the sequences of radioactive changes. For diversion he decided to go to a performance of Wagner's opera *Tristan and Isolde* with his student Oswald Göring. According to Göring, "After a long day of work, Fajans was very tired and soon he fell into a state of somnolence.... I thought that he was asleep, but suddenly he took a piece of paper from his pocket and wrote down an equation.... the development of this equation led to the discovery of hitherto unknown isotopes."[9]

As so often happens, inspiration came with a change of focus and venue. When Fajans entered a dreamlike state and his creative mode emerged, a solution rose to consciousness. After the opera, the displacement laws seemed obvious. All that was needed was a new way of looking at the data.

Fajans, Soddy, Russell, and Hevesy expressed the displacement laws in slightly different ways. In their final form, these laws stated that the electrochemical properties of a decay product depend upon whether the parent element emits an alpha or a beta ray. Alpha ray changes produce a radioelement more electropositive than the parent and shift it through two groups in the periodic table. Beta ray changes yield a product more electronegative than the parent after shifting it through only one group.

Because alpha and beta changes shift the decay product's chemical nature in opposite directions, a combination of three changes, one alpha emission and two beta emissions, will bring the series back to its chemical starting place. The starting element and its third-generation product will be isotopes. "Radioactive children," wrote Soddy later, "frequently resemble their great-grandparents with such complete fidelity that no known means of separating them by chemical analysis exists."[10] The displacement laws were a major breakthrough for radiochemistry. They made sense out of the complex sequences of changes in radioactive decay, as well as the analogies between the uranium, thorium, and actinium decay series.

The simultaneous discovery of the displacement laws created hard feelings. The participants knew each other's research, and some had worked together. Both Soddy and Fajans believed the other stole his ideas. Russell eventually deferred to his mentor Soddy, while Hevesy decided to withdraw all priority claims, citing "the very intricate and for me very awkward situation...." Rutherford wrote to Fajans that "I personally feel that the whole question is a very tangled one, for nearly all the people concerned have talked over the matter with one another.... The consequence is that it is almost impossible without a judge and jury to examine everyone to state the exact origin of the ideas."[11]

What is clear without a court trial is that at the heart of the discovery were the laboratories of Soddy and Rutherford. These pioneer researchers guided and inspired theoretically productive research. The discovery also shows radiochemistry's growing maturity. By 1913 the radiochemical data and analysis had reached the point that researchers of similar training asking the same kinds of questions could reach similar conclusions. "I can honestly say that if Fajans had never existed," wrote Soddy, "it would have made no difference whatever to the whole Periodic Law generalization.... I am quite ready to believe the same could be said by Fajans of Fleck, Russell, & myself."[12]

As with isotopy, Soddy eventually received most of the credit for the displacement laws. One reason may be that he had proposed an incomplete version of the laws in 1911, earlier than the others. More important was his professional status. Unlike the other contenders, Soddy held a professorship, was well known from his earlier researches, and was in a position to reach a wide audience. He wrote well, had done outstanding work, and could express his ideas clearly.

No one questioned the centrality of Soddy's role. But others also deserved recognition. In the end, Soddy benefited from what some historians call "the Matthew effect," after Matthew 13:12: "To anyone who has, more will be given and he will grow rich..." Fame and credit tend to accrue to those who are already famous.[13]

THE END OF THE LINES

As radioelements decayed through the chains beginning with uranium, thorium, or actinium, the process would eventually halt. Researchers no longer would detect rays that signaled the creation of another element. This suggested that the chains ended with inactive products. What could these end products be? Boltwood proposed lead in 1905, citing "the persistent appearance of lead as a constituent of uranium-radium minerals...."[14] Measurements of the lead to uranium ratio in minerals bolstered his hunch. Applying the displacement laws to this problem, both Fajans and Soddy showed that uranium's final descendant indeed should be lead. Subtracting the weight of the alpha particles ejected during uranium's transformations from uranium's atomic weight gave a number slightly under lead's established atomic weight. This meant that lead from uranium minerals should weigh less than ordinary lead.

Scientists rushed to test this prediction in 1914. All found the expected result. Contenders included Otto Hönigschmid, Stephanie Horowitz, Hönigschmid's former teacher Theodore W. Richards, Max Lembert, and Pierre Curie's nephew Maurice Curie.

Hönigschmid was a professor in Prague, then part of Austria-Hungary and the future capital of the Czech Republic. He worked with Horowitz, who came from the Polish region of the Austro-Hungarian Empire. Richards, a renowned Harvard chemist, was an expert on atomic weight measurements. Fajans sent his student Lembert to work with Richards on the atomic weight of lead from uranium. Hönigschmid and Horowitz obtained the most definitive results by finding the atomic weight of lead from radium that was not mixed with other forms of lead.

Soddy and Fajans believed ordinary lead was a mixture of inseparable products from different decay series. Since these products descended from different elements, they should have slightly different atomic weights. Ordinary lead's weight would be an average of the different components. Calculations showed that lead from thorium should weigh more than lead from uranium. Early in 1914 Soddy and his student Henry Hyman determined that lead from a thorium mineral weighed more than ordinary lead. Later, Hönigschmid used a sample that Soddy loaned him to provide highly accurate values for thorium lead's weight.

MORE ISOTOPES

If radioelements and lead could come in more than one form, or isotope, could other elements also have isotopes? The answer came from Cambridge. In 1913 J. J. Thomson was trying to separate ions in a mixture that contained inert gases. To do this, he sent the ions through an arrangement of magnetic and electric fields and recorded their paths with a photographic plate.

Ions from each element in the mix created separate curved lines on the plate, with the position of each curve depending on the atomic weight of the element that created it. Unexpectedly, Thomson obtained a curve which did not correspond to any known element. After considering several possibilities, Thomson decided the line represented a new gas, perhaps a chemical compound of helium and hydrogen.

The strange curve appeared only when neon was present. Thomson's assistant Francis W. Aston tried to separate the particles that created the different curves and measure their densities. One type was heavier than neon's accepted atomic weight, while the other was lighter. Yet their spectra were identical, and matched neon's spectrum. Apparently both types of particles were neon ions, not a chemical compound as Thomson had suggested. Neon's established atomic weight must be an average of the weights of two isotopes. Since neon was not produced by radioactive elements, Aston's results meant that ordinary elements could be composed of isotopes.

In 1919 Aston refined his apparatus to make a device later known as a mass spectrograph. Using this instrument, he showed that many other elements were mixtures of isotopes. For these accomplishments, Aston received the 1922 Nobel Prize for Chemistry.

Inside the Atom

… the atom of a radioactive body is a world …

—Henri Poincaré, 1913

We have here a proof that there is in the atom a fundamental quantity, which increases by regular steps as we pass from one element to the next.... This quantity can only be the charge on the central positive nucleus....

—Henry G.J. Moseley, 1913

BUILDING BLOCKS

The discovery of isotopes shattered one of chemistry's most basic principles: the idea that atomic weight determined chemical properties. If weight was not the key, then what did cause different kinds of atoms to behave so differently? What new idea could make sense of the relations in the periodic table?

The atom was a natural place to look for clues. The problem was that so little was known about its inner workings. *Atom* means "indivisible" in Greek, but by 1900 the term was a misnomer. Physicists no longer envisioned atoms as simple units, but rather as having many parts. Spectra from atoms and molecules were so complex that it seemed only myriads of radiating electrical particles could create them. Most believed these subatomic particles were electrons. The fact that heat, light, and x rays could release electrons from matter, and that radioactive atoms expelled high speed electrons (beta particles) strengthened that assumption.

Since atoms normally are electrically neutral, they must contain a positively charged component that balances the electrons' negative charge. Radioactivity supplied a candidate, the alpha particle. Perhaps atoms were composed of alpha particles and electrons. Some suspected the lightest element, hydrogen, was an ingredient. Ernest Rutherford considered half-helium atoms, and lead (as the final product of the decay series) also was proposed as a building block. The similarities between the three radioactive decay series suggested they shared common structural elements. Early in the twentieth century, several scientists proposed models for the atom's architecture.

BOMBARDING ATOMS

Though no one could see inside an atom, there were ways to get information indirectly. One method was to bombard a substance with electrons or alpha particles. These might strike the object and pass through unchanged, disappear within the object (be absorbed), or bounce back (scatter) and be recorded by a measuring device. By finding the amount of scattered radiation at different angles from the incoming particle's path, physicists could make guesses about the hidden structure that caused their results.

Scattering experiments were popular in Cambridge. J. J. Thomson had pioneered studies of the electron's interactions with matter. He imagined an atom in which multitudes of electrons rotated inside a sphere of positive electricity. Though no one had ever found such a sphere or carrier of positive electricity smaller than an atom, Thomson assumed the atom must contain positive charge. Otherwise, nothing would keep the negatively charged electrons from flying apart.

As Thomson's former student, Rutherford was well prepared for scattering experiments of his own. He was drawn to this subject in 1908 by problems with counting alpha particles, rather than from any plan to uncover atomic structure. Rutherford's assistant Hans Geiger had noticed that scattering was interfering with his counts of alpha particles.

To better understand this scattering, Geiger directed alpha particles through a slit onto a phosphorescent screen, where each particle created a flash. He compared results with the slit uncovered to the outcomes when aluminum or gold foil was placed over it.

Geiger continued these experiments with Ernest Marsden, an undergraduate student in the laboratory. Distribution of the flashes on the screen showed that the scattering's spread depended upon the thickness and composition of the foil. To their amazement, a few alpha particles were scattered by gold foil through angles of 90° or more. How could this be? The most likely scattering angle, Geiger calculated, was only about 1°. "It was as though," Rutherford explained later, "you had fired a fifteen-inch shell at a piece of tissue paper and it had bounced back and hit you."[1]

To understand scattering quantitatively, a mathematical theory was needed. In 1910 Thomson published such a theory based on his atomic model. Thomson envisioned an atom with electrons distributed throughout it, without being concentrated in any particular region. Scattering experiments at the Cavendish Laboratory by James A. Crowther seemed to agree with Thomson's model. However, this model did not account for the large-angle scattering that Rutherford's students observed.

Ambitious and confident, Rutherford could barely hide his desire to outdo his scientific elders. In 1904 he had undermined the venerable Lord Kelvin's predictions about the earth's heat. Now he took on Thomson by jousting with Crowther. Rutherford spent months tinkering with scattering formulae and criticizing Crowther's work privately. Finally, he devised a model that seemed to work.

THE NUCLEAR ATOM

The only way to explain large-angle scattering, he reasoned, was to suppose that an alpha particle occasionally came very close to a massive charge that kicked it almost straight back. Since this event was so unusual, the atom must be very sparsely populated, so that most projectiles passed through with little or no interference.

Rutherford decided that the atom's positive charge must be concentrated in a small volume (a nucleus) in order to pack such a punch. He imagined an atom that looked like a miniature solar system, with rings of electrons rotating around a positively charged nucleus. He assumed the charge on the nucleus was an integer since its possible components (alpha particles and electrons) carried whole-number charges. Rutherford then developed equations for scattering based on this model.

Geiger ran experiments to test these equations. The results agreed well with predictions. Rutherford published his theory in 1911, but as a theory of scattering rather than as an atomic model.

Looking back, this seems a momentous event. After years of speculation about the elusive atom, something definite was taking shape: a rarefied atom with a highly charged central core. But even Rutherford seemed not to realize his discovery's significance in 1911. His aim was to develop an improved scattering theory, not to explain atomic structure. The nucleus was something of an afterthought, and neither he nor his contemporaries stressed it.

Rutherford's atomic model was not even original, as the Japanese physicist Hantaro Nagaoka had proposed a solar system-type model in 1906 and Jean Perrin had suggested something similar in 1901. Philipp Lenard and William H. Bragg had long recognized that the atom must be mostly empty space. The idea of an atomic nucleus or central core was commonplace, making it easy to overlook this feature of Rutherford's paper.

Geiger and Marsden's experimental data and Rutherford's analysis were not available to the early model builders. Since Rutherford's theory could be quantified and tested, it could be developed further than the earlier speculations. The wan interest in his 1911 publication proved to be temporary. Prospects for Rutherford's nuclear atom changed radically after a young Danish physicist named Niels Bohr arrived in Manchester.

THE NUCLEUS AND THE PERIODIC TABLE

Son of a physiology professor at Denmark's University of Copenhagen, Bohr studied physics and wrote a dissertation on electrons in metals. He went to Cambridge hoping to get it published and to convince Thomson that his theory was erroneous. After a couple of months, Bohr realized he was not getting anywhere with this project. He decided to apply for a place in Rutherford's laboratory.

The earnest Dane managed to impress Rutherford in spite of his long-windedness. Among Bohr's colleagues in the laboratory were Georg von Hevesy and Alexander Russell, who discussed the puzzles of the radioelements and their decay sequences with Bohr as they worked out the displacement laws, and two young physicists, Charles G. Darwin, grandson of the naturalist Charles Darwin, and Henry G. J. Moseley.

One idea these scientists discussed was an alluring hypothesis that lacked experimental proof. In 1911 Antonius J. van den Broek, a Dutch lawyer who closely followed developments in recent physics, made a simple yet profound conceptual leap. Scientists had been ordering the elements in the periodic table according to their atomic weights. Isotopes were a problem for this system, since isotopes of a single element had different atomic weights. Van den Broek perceived that it might be useful to characterize the elements by their positions in a sequence, instead of by their atomic weights. The number sequence would begin with the first element, hydrogen, and end with the ninety-second element, uranium.[2]

The next year van den Broek proposed that these numbers represented each element's nuclear charge. Hydrogen would have a charge of 1, uranium 92, and so forth. Rather than atomic weight, these integers would determine the elements' chemical properties. They became known as *atomic numbers.*

Without experiments to support it, van den Broek's idea was just an interesting speculation. Evidence came from x rays emitted from metals bombarded by high speed cathode rays. Around 1906 George Barkla, a former student of J. J. Thomson, reported that elements he had bombarded produced patterns of x rays specific to each element. These so-called characteristic x rays seemed to come from deep within the atom.

In 1912, after years of controversy, experiments finally showed that x rays were a type of electromagnetic radiation similar to light. The next year William H. Bragg and his son William Lawrence Bragg, working at the University of Leeds in the north of England, discovered a way to determine the wavelengths (and frequencies) of x rays. Moseley and Darwin eagerly began experiments with x rays at Cambridge. After they finished a paper together, Moseley decided to use the Braggs' method to test van den Broek's hypothesis. He would measure the frequencies of characteristic x rays from as many elements as possible and see if these frequencies were related to atomic numbers.

Moseley obtained purified samples of twenty-seven elements, which he pummelled with cathode rays. He measured the frequencies of the characteristic x rays that resulted. As predicted, Moseley found a mathematical relationship between the x-ray frequencies and the elements' atomic numbers. This was strong support for van den Broek's hypothesis.

Van den Broek had claimed that atomic number corresponded to nuclear charge. Some scientists had already inferred this and others were ready to accept it. Bohr and others realized that shifting their focus from weight to atomic number would solve the problem the radioelements posed for the periodic table. By using atomic number to sequence the elements, all of an element's isotopes would fit into one position in the table. The fact that their atomic weights differed did not matter. Soon atomic number became the periodic table's new organizing principle.

Bohr was keenly interested in these problems. He wanted to build a theory of the atom which combined current knowledge with the new quantum theory proposed by Max Planck in 1900. According to this theory, energy is transferred only in specific quantities, or quanta, rather than being transferred in any amount. After a few months in Manchester Bohr returned to Denmark to marry his fiancée Margrethe Nørlund and to take a position at University of Copenhagen. In 1913 he published a lengthy three-part paper on the structure of atoms and molecules. Using Rutherford's nuclear atom and van den Broek's scheme, Bohr assigned each nucleus a charge proportional to its numerical position in the periodic table (atomic number). He proposed that atomic number represented not only the nuclear charge but also the number of outer electrons in an atom. The negative charge of the orbiting electrons would balance the positive charge of the nucleus.

Bohr incorporated the quantum into his atomic model by making bold assumptions to simplify the mathematical problems involved. His model worked reasonably well for the simplest atom, hydrogen, but failed miserably for more complex atoms. The theory's difficulties seemed intractable. Yet, Bohr managed to explain long-standing puzzles in spectroscopy with his theory. Many found his theory attractive because it accounted for a variety of phenomena and combined new ideas in a harmonious way. It took over a decade of work by Bohr and others and major revisions to the theory, but Bohr's ideas eventually became part of mainstream physics.

Interest in Rutherford's nuclear atom received a boost from Bohr's elaboration of this model. In retrospect, the model's success was assured. Over a decade of research had already prepared the way for the nuclear atom. Radioactivity's stubborn resistance to any attempts to control it had convinced scientists that it came from a part of the atom different from the region of ordinary chemical and physical effects. The nuclear atom fit this requirement well, and by 1914 Rutherford's model was widely accepted.

THE GAMMA RAYS

How did the gamma rays fit into the radioactive atom? To answer this question, scientists would need to know more about these invisible rays. Though the gamma rays seemed to be highly penetrating x rays, researchers considered other possibilities. In 1902 Rutherford noticed that, in some ways, they behaved more like cathode rays, which were known to be electrons. Robert J. Strutt also wondered whether the gammas might be particles.

Several other physicists took up this idea, including the German physicist Friedrich Paschen and William H. Bragg. Experiments eventually showed that the gamma rays carried no charge, so they could not be electrons. Perhaps the gammas were uncharged particles. Bragg proposed that they were particles formed from a positive and a negative charge. These paired charges would create a neutral particle, which he called a "neutral pair."[3]

The idea that the gamma rays were particles was based on experimental facts. Some results with gamma rays, x rays, and light were hard to explain without assuming these rays were concentrated in space, rather than spread out like a wave or a pulse of energy. One example was the photoelectric effect, where light knocks electrons out of atoms. The electrons can be collected as a current that can trigger a switch, for instance, to open a door when a person intercepts a light beam. It was hard to imagine how enough energy could be transferred to the electrons if it were spread out over space like a wave or a pulse (an abbreviated wave) of energy, but easy to visualize if the energy were concentrated in a particle. Bragg believed that his corpuscular model fit the behavior of these radiations better than the wave and pulse models.

Publication of Bragg's neutral pair theory touched off a controversy with Barkla. Their protracted argument led to important findings, including Barkla's discovery of the characteristic x rays. Many others entered into the debate on the nature of the gamma rays, including John Madsen in Australia; J. J. Thomson in England; Wilhelm Wien, Johannes Stark, Edgar Meyer, and Arnold Sommerfeld in Germany; Egon von Schweidler in Austria; and Frederick Soddy in Scotland.

In 1912 two physicists at the University of Munich, Walter Friedrich and C. M. Paul Knipping, showed that x rays reflected from crystals behaved like waves. They based their experiments on a theory of Max von Laue, who was then lecturing at the university. (Laue received the 1915 Nobel Prize for his groundbreaking work.) After learning of the 1912 experiments, Rutherford and Edward Neville da Costa Andrade, who was in the Manchester laboratory on a fellowship, decided to try these experiments with gamma rays.

The gamma rays bounced off rock salt crystals like x rays, allowing Rutherford and Andrade to measure some gamma wavelengths. For most scientists, these experiments proved the wave nature of the gamma rays.

Yet the contradictions remained, and Bragg did not abandon his neutral pair theory. It seemed to him, as it had to Soddy, that both the wave and corpuscular theories were correct. The paradox simply had to be accepted. "The energy travels from point to point like a corpuscle; the disposition of the lines of travel is governed by a wave theory: Seems pretty hard to explain: but that is surely how it stands at present."[4] Bragg believed the quantum theory would successfully combine the opposing wave and corpuscle models. This might work mathematically, but a theory that combined such different models would be difficult to visualize. Most of Bragg's British peers were not convinced they needed to take such a radical step.

Work during the 1920s eventually confirmed Bragg's hunch. Both the wave and particle models were correct. This paradox became enshrined in physics as wave-particle duality, a fundamental principle of physics. According to this principle, both models are needed to describe the natural world.

THEORIES OF THE NUCLEUS

Around 1912 researchers began talking about different kinds of electrons. Those near the atom's surface governed electrical conduction, spectra, the photoelectric effect, thermal radiation, and chemical behavior. Next came the electrons whose vibrations created the characteristic x rays. Radioactivity's source was even deeper inside the atom. Inaccessible to the experimenter, radioactivity revealed itself only by mysterious explosions that hurled alpha or beta particles or gamma rays out of the atom. The nucleus seemed a likely site for radioactivity.

The alpha and beta particles came either from the nucleus itself or from some place very close to it. Since these particles were so energetic, scientists supposed they were already moving before they were released from the atom. Most assumed the nucleus contained alpha particles, which gave it a positive charge, and that the beta particle electrons were arranged in rings, like the outer electrons. Rutherford thought the beta ray rings were outside the nucleus but deeper inside the atom than the rings that produced the characteristic x rays. Van den Broek envisioned beta rays coming from electron rings rotating inside the nucleus itself.

Many scientists assumed that beta particles generated gamma rays as they burst out of the atom. Rutherford thought the beta rays came from the atom's innermost rings and excited gamma rays when they passed through the outer electron rings. Others suspected that both the beta and gamma rays came from the nucleus.

By 1913 evidence pointed towards a nuclear origin for the gamma rays. Rutherford modified his model by moving the primary beta particles to the nucleus, but he did not relocate the gamma rays. He still believed some beta particles came from electron rings outside the nucleus. As the primary beta particles shot out of the nucleus, they would knock the secondary beta particles out of the electron rings and produce gamma rays.

Nuclear electrons could provide an explanation for isotopes. If the nucleus contained both electrons and alpha particles, nuclei with different numbers of alpha particles could carry the same net charge and atomic number. For instance, a nucleus containing four alpha particles would have a charge of positive 8, the same as a nucleus containing five alpha particles (positive 10) and two electrons (negative 2). Both nuclei would have the same atomic number, 8, but different atomic weights. They would be isotopes.

Nuclear electrons could also furnish a sort of glue to hold the alpha particles together in the nucleus. Without some type of binding force, mutual repulsion would drive these positively charged particles apart. To avoid this difficulty, some scientists suggested that attractive and repulsive electrical forces did not exist inside the nucleus, or that other kinds of forces held the alpha particles together.

Several scientists proposed nuclear models for radioactivity in 1915, including John Nicholson and Frederick Lindemann in England, André Debierne in France, Heinrich Rausch von Traubenberg in Germany, and William D. Harkins and Ernest D. Wilson in the United States. Debierne introduced the prescient idea of a nucleus constrained by surface tension, which was reinvented in the 1930s and used to explain nuclear fission.

However tantalizing these ideas seemed, scientists had no way to test them. The nucleus and its forces were unknown territory until well after World War I.

Sequel

I would advise England to watch Dr. Rutherford....Great developments are likely to transpire shortly...

—*Marie Curie*, 1913

WAR!

By 1913, the world had changed dramatically in the eighteen years since radioactivity's discovery. The Wright brothers had flown an airplane, and Henry Ford was producing the model T automobile. Storefronts featured neon signs, while people sent messages by telegraph and telephone. Chemistry had produced amazing new materials, such as bakelite, cellophane, and rayon. Biochemists were closer to understanding vitamins, enzymes, and hormones and had deciphered chlorophyll's composition.

The globe was yielding to explorers and archaeologists, who conquered the South Pole and the Arctic and uncovered a lost civilization at Knossos in Crete. Technological giants like General Motors, Ford, Daimler, and I. G. Farben were formed during this period, as well as cultural institutions like Dublin's Abbey Theatre and Munich's Deutsches Museum. More women were entering universities, many of which had only recently opened their doors to female students, and suffragettes were demonstrating in London. In the domestic sphere, parents bought teddy bears for their children and read them Beatrix Potter's storybooks.

This had not been a peaceful time for Europeans, who had waged wars in South Africa, East Asia, and Italy. The Balkan region, long simmering with cultural and political tensions, flashed out in a series of conflicts which ended with uneasy armistices during 1913.

Art reflected the changing times. Innovators like Picasso, Matisse, Chagal, Derain, and Kandinsky demonstrated new ways of seeing the world with cubism, fauvism, German expressionism, and other styles known collectively as post-impressionism. The 1913 Armory show in New York featuring the new approaches was a sensation. Always on the cutting edge of culture, the art world's move away from realistic depictions of ideal subjects foreshadowed the jarring unreality of future events.

In 1914 the fragile European détente disintegrated, with unimaginable consequences. The provocateur was a Serbian nationalist who assassinated Archduke Francis Ferdinand of the Austro-Hungarian Empire. This act triggered an escalating chain of events that exploded into World War I.

Ordinary life, including most scientific research, came to a standstill. Of the major laboratories for radioactivity, only the Vienna Radium Institute managed to keep research going throughout the war.[1] Some scientists, like Ernest Rutherford, worked on technical projects for the military, while others, like Otto Hahn, were drafted. Marie Curie and her seventeen-year-old daughter Irène drove around French battlefields to bring traveling x-ray stations to the front. Lise Meitner, an Austrian physicist working in Berlin, volunteered for the Austrian army as an x-ray nurse-technician.

A few unfortunates, like Cambridge physicist James Chadwick who was working with Hans Geiger in Berlin and Rutherford's glassblower Otto Baumbach in Manchester, were caught on the wrong side of the battle lines and spent the war in captivity.[2] Many promising researchers lost their lives, including Henry G. J. Moseley and Marie Curie's protegé Jan Danysz.

World War I turned out to be the most devastating conflict known to humanity. Science and technology were enlisted to create new weapons, including poisonous gases, and new ways to fight using airplanes and submarines. Europe bore physical and psychological scars for generations.

After the defeat of Germany and Austria brought the war to an end, researchers gradually resumed work. Conditions were particularly difficult for the war's losers, who faced shortages of food and heating fuel. Deadly influenza was rampant, while social unrest and lawlessness flourished. Harsh terms imposed by the victors increased the suffering and humiliation, making Germany especially vulnerable to an authoritarian movement. This dire outcome materialized in the 1930s with the takeover of Germany by the National Socialist German Workers' Party, otherwise known as the "Nazis."

RADIOACTIVITY DURING WORLD WAR I

In spite of the draft and diversion of laboratories and funds, some scientists were able to work on radioactivity during the war. Searches for missing links in the radioactive decay series, measurements of decay periods and atomic weights, behavior of radiations, relations between decay periods and alpha particle ranges, luminous paints, and radioactive dating were typical topics pursued.

Though many details awaited completion, the overall scheme of the radioelements, their characteristics, and their interrelationships were nearly complete by the outbreak of the war. After reading Stefan Meyer and Egon von Schweidler's 1916 textbook on radioactivity, one reviewer concluded that this science was coming to a close.

Predictions of a field's demise had been made before. Towards the end of the nineteenth century, after the successes of Newtonian mechanics, electromagnetic theory, and thermodynamics, some thought that physics left little for the researcher to uncover.

Unforeseen discoveries turned that prophecy on its head. X rays, radioactivity, the electron, and other surprises shattered any notion of physics' completeness and opened up a wealth of new questions to explore.

Despite a superficial impression that radioactivity would soon be a closed book, those closest to the field knew the most basic questions were still open. How did the atom generate such huge amounts of energy? What determined which atoms would decay at a given time? How could a causal explanation account for the random nature of decay? What linked the rays and atomic structure?

Attempts were made before and during the war to confront these riddles, yet progress seemed blocked. Tasks like making better measurements and clarifying the genetic relations between radioelements held better chances of success, particularly with wartime shortages of time, resources, and personnel. In retrospect, researchers were missing crucial information needed to unravel the persistent puzzles. That knowledge did not become available until the late 1920s and early 1930s, with advances in quantum theory and the advent of wave mechanics.

Some research begun during the war opened up exciting possibilities, burying any rumors that radioactivity's potential was exhausted. Ironically, these findings, combined with other circumstances, led radioactivity down a path which ended its existence as a separate research field.

FROM RADIOACTIVITY TO NUCLEAR AND PARTICLE PHYSICS

Right before the war Rutherford's student Ernest Marsden reported a tantalizing result. While experimenting with alpha particles, he noticed high speed hydrogen atoms in the apparatus. Perhaps hydrogen, like helium, was a by-product of radioactive disintegration.

This possibility intrigued Rutherford. Marsden left Cambridge and could not continue experiments while he served in the army. War research claimed most of Rutherford's time, but he managed to continue Marsden's investigations sporadically. To his surprise, Rutherford found that radioactive decay was not responsible for hydrogen's appearance. The hydrogen in Marsden's apparatus came from nitrogen atoms in the air.

Apparently an energetic alpha particle could chip off a piece of a nitrogen atom, producing hydrogen atoms. After years of vain attempts to influence radioactive disintegration, Rutherford realized that he and Marsden had accidentally caused an ordinary element to disintegrate![3]

If alpha particles could split off hydrogen atoms from other atoms, hydrogen might be an atomic building block. These results bolstered long-standing suspicions about the atom's composition.[4] They also presented an exciting possibility. Perhaps powerful alpha particles could be used to disrupt other atoms and uncover information about nuclear forces and structure.

Rutherford's results inspired new research with alpha and beta particles and efforts to get more powerful projectiles for bombarding atoms. To increase the energies of these subatomic bullets, scientists invented machines that could accelerate particles to high speeds. The electrostatic generator, the linear accelerator, and the cyclotron developed in the 1920s and 1930s were the first in a series of increasingly bigger and more ingenious devices for smashing atoms.

Several important textbooks on radioactivity appeared in the 1920s. Laboratories in Paris and Vienna published scores of papers on radioactivity researches, while sites like Berlin and Cambridge were not far behind. Meyer and von Schweidler cited 1,561 authors for the period 1916–26 in their textbook *Radioaktivität*. The field was healthy, but interests were changing.

Rutherford assessed the discipline in his 1930 textbook, *Radiations from Radioactive Substances*. The title embodies the change he recognized, a shift from studying radioactive transformations and decay relationships to understanding the rays and their interactions with matter. These studies, he believed, would reveal information about two persistent puzzles in radioactivity: the structure of the nucleus and the energy changes involved in its transformations. Bombarding matter with high speed particles was the new frontier of research.

During the 1920s physicists had increasingly directed their attention towards the atomic nucleus and the subatomic particles resident in the universe. Theoreticians became interested in the previously experimental science of the atom. They applied quantum theory and wave mechanics to the nucleus. By delineating probability's reign in the physics of the atom, the new theories placed chance at the heart of nature. Though some physicists still desired causal explanations for radioactive decay and the quantum, they were willing to move beyond those seemingly impenetrable mysteries to questions they could answer.

In 1931 the first International Conference on Nuclear Physics convened in Rome. The next year Chadwick identified an uncharged subatomic particle, the neutron. Of significant theoretical import, the neutron became another tool for probing the atom's interior. A positively charged counterpart to the electron, the positron, was discovered in cosmic rays by Carl Anderson.

In 1934 Irène Curie and her husband Frédéric Joliot produced radioactive phosphorus, the first of a sequence of artificially created radioactive isotopes. Many of these isotopes became useful for medicine and industry, for instance as radioactive tracers, for treating cancer, and as ionization sources in smoke detectors.

Experiments with artificial isotopes eventually led to the discovery of nuclear fission in 1939 by German chemists Otto Hahn and Fritz Strassmann and Austrian physicist Lise Meitner. In fission, an unstable nucleus breaks apart, producing two atoms of lesser atomic weight. Commonly known as "splitting the atom," this process releases huge amounts of energy.

Without announcement or fanfare, radioactivity had been superseded by nuclear physics and particle physics. That once fascinating phenomenon remained interesting mainly for its applications. Another new field, nuclear chemistry, took the place vacated by radiochemistry, whose major problems had been solved by the early 1920s. Cosmic ray studies, born from radioactivity research, were swallowed up by particle physics.

And so the mystery which had enticed scientists for several decades was never solved on its original terms. New ways of conceptualizing physics replaced the older ideas of physical systems operating through chains of cause and effect. Successes of the new approaches convinced most scientists to view probability as a basic

principle of the physical world requiring no further explanation. The mystery of radioactivity was replaced by a new set of questions about nuclear structure, particles and fields, quantum transitions, symmetry and invariance, and increasingly abstruse mathematical schemes. The solidly experimental science of radioactivity transmuted into new theoretical species, nurtured by their experimental offshoots and trailed by a series of practical consequences.

PART TWO

MEASURING AND USING RADIOACTIVITY

And now you are in possession of salts of pure radium! ... What a pity it is that this work has only theoretical interest, as it seems.
—*Władisław Skłodowski to Marie Curie, 1902*

Circumstances soon superseded these words, written by Marie Curie's beloved father only a few days before his death. Radioactivity's potential for practical applications fueled a burgeoning industry for radioactive materials. This new industry had far-reaching effects.

Radiation's visible effects on living tissues stimulated searches for medical uses, with unforeseen consequences. Demand for radioactive materials spurred widespread exploration and business ventures to process and manufacture radioactive materials. New instruments opened up fresh avenues for exploring the unknown, as well as more options for practical applications of radioactivity.

International standards for radioactivity facilitated research, medicine, and industry. Measuring methods and instruments grew more sophisticated and eventually transformed the scale of research. Research, medicine, and industry became entangled with social, political, and ethical questions raised by the gradual recognition of radioactivity's powers.

Methods and Instruments

Things have their due measure.

—Horace, *Satires*

CRUCIAL CHOICES

When scientists select a particular method to investigate a phenomenon, they implicitly accept underlying assumptions about the phenomenon and narrow the range of options to consider. The ways that experimenters gather and record data determine what they can observe and limit the possibilities for interpreting their results. Methods chosen to investigate radioactivity influenced the growth and direction of the new science.

Early experimenters could use photography, electrical measuring devices, and fluorescent minerals to study radioactivity. Each method had particular strengths and limitations, and each suggested a different interpretation of radioactivity. To employ photography implied the new rays were similar to light. Henri Becquerel used photography for his first experiments because he was searching for a new kind of light. Since electrical methods record ionization, they could suggest a resemblance to x rays, cathode rays, or ultraviolet light. Testing with fluorescent materials implicated ultraviolet light or cathode rays.

Photography was familiar and simple in concept. Photographs captured direct visual images, useful for qualitative studies and for recording the paths of rays deflected by electric and magnetic fields. Unfortunately, this method had serious drawbacks. Photographic images were difficult to quantify and interpret. For instance, it was hard to determine intensity of radiation from photographic impressions. Photography could not be used for short-lived radioactive products because longer exposure times were needed to produce useful images. Photography is especially prone to error, because many agents besides radioactivity can affect photographic plates and film.

Scientists used electrical devices to study ionization produced by radioactive bodies. Fitted with scales to measure the movement of electroscope leaves or electrometer needles and strings, electroscopes and electrometers could be extremely sensitive and yield precise quantitative values. According to Pierre Curie, his electroscope was 10,000 times more sensitive than the spectroscope, an instrument which could detect vanishingly small quantities of an element.[1]

Since electrical instruments responded readily and could be easily modified, they could be adapted for all strengths and types of radiations. Unlike photographic plates, electrical devices could detect short-lived radioactive bodies. Because of their advantages, electrical methods soon displaced photography in radioactivity research.

Becquerel rays create striking visual effects in fluorescent minerals, making them candidates for detecting radioactivity. Nineteenth-century researchers had used screens made of fluorescent barium platinum cyanide to detect rays. These screens became popular for studying cathode rays and x rays and were later used to investigate Becquerel rays. This simple and colorful method was not suited for most quantitative studies.

After Ernest Rutherford and Hans Geiger showed in 1908 that alpha particles create individual flashes of light when they strike fluorescent zinc sulfide (Sidot's blende), fluorescent screens could be used for quantitative work. By counting scintillations, researchers were actually counting alpha particles. Zinc sulfide screens became widely used for this purpose.[2]

Since alpha, beta, and gamma rays affect photographic plates, fluorescent materials, and electrical devices differently, the instrument chosen determined what was recorded and could influence the conclusions drawn from data. Other details of experimental arrangements might also interfere with interpretations.

For instance, paper or glass coverings on photographic plates allowed beta and gamma rays to pass through, but blocked weak alpha rays. When researchers separated uranium and thorium from their beta-emitting decay products in 1901, they found no activity in the uranium and thorium. Sir William Crookes, Becquerel, Rutherford, and Frederick Soddy mistakenly concluded that these elements were not radioactive. Only later did they learn that uranium and thorium sent out alpha rays. Since the alpha rays could not pass through the paper or glass coverings in the earlier experiments, they had not been detected.

STANDARDIZING THE MEASURES

At first, laboratories developed their own methods for measuring radioactivity and determining the amount of radioactive material in preparations. As research proliferated, it became increasingly desirable to find a universal method for comparing these measurements. Researchers, physicians, and manufacturers needed to know the amount of radium in a sample in order to compare research results, measure out quantities for therapy, and sell radium in an international market.

One popular way of comparing the radioactivity of samples was to compare the amount of emanation they emitted. A more convenient method involved measuring gamma radiation from radium preparations. This method avoided the difficulties of working with gases, and since gamma rays could pass through glass, samples could be safely sealed. Radium and radium emanation were usually chosen for standards, and other radioactive substances were compared to them. When mesothorium became commercially important, Otto Hahn developed his own standards for comparing mesothorium preparations in order to price them correctly. Other methods were developed later for substances that did not emit gamma rays.

In 1910 ten prominent radioactivity researchers gathered in Belgium to develop standards for radioactivity: Marie Curie and André Debierne from France, Hans Geitel and Hahn from Germany, Stefan Meyer and Egon von Schweidler from Austria, Rutherford and Soddy from Britain, Arthur S. Eve from Canada, and Bertram B. Boltwood from the United States. Known as the International Radium Standards Committee, the group negotiated methods, nomenclature, and the best place to store a radium standard. They decided to name radioactivity's basic unit of measurement the "curie," defining it as the amount of emanation in equilibrium with one gram of radium.

In order to develop a standard quantity of radium for comparing measurements, Marie Curie and atomic weight expert Otto Hönigschmid prepared samples of purified radium chloride. The committee met again in 1912 to compare these specimens. After finding agreement between Curie's and Hönigschmid's results for the amounts of radium in their samples, the committee decided (in deference to Curie) to adopt Curie's sample as the international standard.

Curie's 21.99 milligrams of radium chloride, sealed in a glass tube, was transferred to the International Bureau of Weights and Measures near Paris. One of Hönigschmid's preparations remained in Vienna as a secondary standard. Creation of the international standard facilitated accurate, reproducible work in science, medicine, and industry. The committee arranged to provide duplicate standards for governments that requested them. By 1925 duplicate standards were stationed in Paris, Brussels, Middlesex (England), London, Washington (D.C.), Vienna, and Berlin.

INNOVATIONS

The new science's progress depended upon instruments and techniques developed during the nineteenth century. Electrical research received a huge boost from several mid-century innovations: the Geissler tube, the Rühmkorff induction coil, and the Sprengel pump. Experiments using Johann Geissler's tubes evacuated with Hermann Sprengel's pump and powered by Heinrich Rühmkorff's induction coil revealed cathode rays and x rays. The hunt for x rays in turn led Becquerel to radioactivity. Photography, another nineteenth-century invention, enabled Becquerel to make the discovery.

Radioactivity researchers began with the available instruments. Becquerel, who believed uranium rays were a form of light, captured his results on photographic plates. The Curies and

Rutherford focused on radioactivity's electrical effects and used electroscopes or electrometers. Over time, researchers adapted and modified their instruments and techniques. They made these devices more convenient to use and better suited for recording radioactivity's effects. To analyze radioactivity, researchers harnessed new theories of ionization and conduction, which were based on experiments using electrical instruments.

The new field soon inspired new instruments. Fluorescent zinc sulfide screens first marketed by Friedrich Giesel were used widely to detect alpha rays and later to count alpha particles by observing the scintillations they produced.

The spinthariscope (from Greek words meaning "spark" and "to view") was a convenient tool for viewing scintillations. Invented by Crookes, it soon became a popular novelty. The spinthariscope consisted of a handheld tube with a magnifying eyepiece on one end, a zinc sulfide screen on the opposite end, and a small amount of radium. When alpha particles from the radium struck the screen, they created scintillations, which could be viewed through the eyepiece and counted.

This pocket-sized device made atom-sized events visible, as each scintillation signaled the arrival of a single alpha particle. It convinced at least one famous physicist (Ernst Mach) of the atom's reality.[3]

Later, researchers used photoelectric devices to detect alpha particle scintillations. In these instruments, the flash of light from an alpha particle collision ejects electrons from a light-sensitive material. The electrons travel to an electrical measuring device that records them. The modern photoelectric "eye" used to open and close doors works on the same principle.

Weak ionization was difficult to detect with ordinary electrometers. The counting device first built by Geiger in 1908 provided an alternative. This instrument could be modified to count beta, gamma, or cosmic rays. Geiger worked with Walther Müller in 1928 to develop an especially sensitive model, known as the Geiger-Müller counter.

Charles Thomson Rees Wilson (known as C.T.R.) invented another new instrument to record rays. Because this device works on the same principles that rainmakers use to make artificial clouds, it was called the "cloud chamber."

Clouds are made of raindrops formed when water vapor condenses on particles in the air. Wilson began investigating clouds in J. J. Thomson's laboratory in 1896. He found that he could create them by ionizing air with x rays or uranium rays. Water vapor could condense on these small particles and form a cloud along the path of the rays. For the water vapor to condense, the air must be cooled, which Wilson did by allowing the air in his apparatus to expand suddenly. A beam of light made the cloud visible. In 1897 Wilson obtained a rough value for the electron's charge by observing the droplets that formed in his cloud chamber when he ionized the air inside it. Assuming that each charged particle created one droplet, he divided the total charge on the droplets by the number of droplets to obtain the charge on one particle.

Wilson's research sprang into the spotlight in 1911, when he found that a single alpha or beta particle created a trail of droplets that he could photograph. These tracks showed vividly that electrons and atoms (or ions like the alpha particle) were more than convenient models—they were real objects.

Using the cloud chamber, scientists could trace movements of individual invisible particles. They could measure a particle's range from its tracks and compute its energy. Tracks in a cloud chamber could record collisions between particles and provide information about them. Cloud chambers became very useful in the 1920s and 1930s for studying cosmic rays and in nuclear and particle physics.

Another method for recording particle tracks was introduced by Japanese physicist Suckichi Kinoshita (in 1910) and German physicist Maximilian Reinganum (in 1911). These researchers showed that alpha particles made tracks in the coatings (emulsions) used on photographic plates. Soon new emulsions made more sophisticated experiments possible. During the 1920s and 1930s scientists used photographic plates to record cosmic ray particle tracks. Plates were easier to transport and to use than the cloud chamber. They were always ready to record an event, unlike the cloud chamber which needed to be set and reset each time a particle entered the instrument.

Besides using photography to record ray tracks, scientists used this method to create images of radioactive objects they placed on photographic plates. These images were called "autoradiographs" (meaning self-radiograph). An autoradiograph is the image that an object's own radiation creates on a plate or film, while a radiograph is the image of an object's shadow, created when an outside source (like x rays) strikes the object.

The first autoradiograph, an image of a uranium sample, led Becquerel to radioactivity. During the 1920s medical and biological researchers began using autoradiographs to show the distribution of radioactive substances injected into plant and animal tissues. Autoradiography became an important tool for understanding how different substances travel in living things and where they concentrate in organs and tissues. Investigators also used this technique to trace movements of metals in alloys subjected to heating, cooling, and other changes. With this information, they could recommend improvements for alloy manufacturing.

In addition to devices that were widely used and marketed, laboratory technicians constructed complex apparatus essential for the experimenters who investigated radioactivity. Without the skill, inventiveness, and perseverance of mechanics, glass blowers, demonstrators, laboratory stewards, and other assistants, many investigations would not have been possible.

SIZE, MONEY, AND MACHINES

Aside from expensive radioactive materials, the earliest research required rather simple resources. Scientists could do significant work outside a traditional laboratory, for instance at home or in the field, or in straitened circumstances. The story of the discoveries of radium and polonium in a dilapidated shed, the research in Vienna in an antiquated building, Lise Meitner working in a former woodshop in Berlin, and the anecdotes about Rutherford improvising with tin foil and coffee cans are legendary. According to a colleague, Rutherford once claimed he could do research at the North Pole.[4]

Rutherford's penchant for improvising continued after he moved to the Cavendish Laboratory at Cambridge in 1919. When Hungarian physicist Elizabeth Rona first entered his basement laboratory, she noticed that "the instruments looked primitive, self-made." Rather than being concerned about the widespread radioactive contamination, her first fear was electrocution from low-hanging high voltage wires (Figure 9-1). After being shown a famous piece of equipment (the Cockcroft-Walton particle accelerator), a journalist wrote that it looked like it was constructed from "plasticine, biscuit tins, and, I suspect, sugar crates."[5]

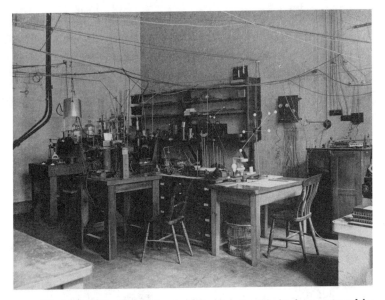

Figure 9-1. Rutherford's laboratory, early 1920s. Reproduced by kind permission of the Syndics of Cambridge University Library, MS.Add.7653:PA336

In spite of these tales, scientists coveted advanced facilities and equipment. They designed accurate and sensitive devices and purchased the best instruments they could afford. Costly apparatus like strong electromagnets for studying the rays and low temperature equipment to liquify radioactive gases placed their possessors at an advantage. Giesel, who had factory resources from the beginning, made important contributions in spite of his limited time for research. McGill University lured Rutherford with its state-of-the-art facilities, and Cambridge University later enticed him with its famed Cavendish Laboratory. The Curies continually pressed for improved facilities. Elster and Geitel, who did not have the means of the larger centers, focused on questions they could pursue in the field and in their home laboratory.

The days when science could be pursued on such a small scale were numbered. Even before World War I large laboratories were dominating the field, though pathbreaking research was still being done in more modest facilities. During the 1920s funding and resources became increasingly centralized, along national lines.

As radioactivity gradually transformed itself into nuclear physics, the scale of research changed dramatically. Scientific centers in Berlin, Rome, and Berkeley rose to prominence in the 1930s. Nationalism continued to spur research, miscarrying it into a disastrous war. World War II and its "cold war" aftermath raised the stakes and scale. Escalation towards bigger and bigger physics led to gargantuan particle accelerators and astronomical costs. The process eventually reached a limit in the United States towards the end of the twentieth century, with the demise of the superconducting supercollider project in 1993. Still, this was not the end of the supersized machines. The Conseil Éuropéen pour la Recherche Nucléaire's (CERN's) Large Hadron Collider in Geneva, with a seventeen mile underground circular path to accelerate particles, began operating in 2009.

Radioactivity, Medicine, and Life

We always handled our preparations with our bare hands and stirred them around with our fingers.

—*Otto Hahn*, 1968

The challenge of unraveling a new discovery was irresistible to curious scientists, who required no further justification for investigating radioactivity. As the enterprising began to envision uses for radioactivity, they encouraged research with more pragmatic ends. Most compelling was radioactivity's potential for medicine.

UNPLEASANT SURPRISES

The first hints that radioactivity affected living creatures emerged a few years after the discovery. In 1900 Friedrich Giesel loaned one-fifth of a gram of radium to his friend Friedrich O. Walkoff, a dentist who wanted to study radium's effects on the human body. Walkoff made his first test on himself. After forty minutes' exposure to radium, the skin on his arm became inflamed. This lesion did not heal readily. Giesel then experimented with his own arm, leaving the radium in place for two hours. Not only did his skin turn red; the longer exposure caused it to peel. When Giesel sent a radium sample to spectroscopist Carl Runge, he warned Runge of radium's "most unpleasant physiological effect … if even the tiniest amount of the substance comes into contact with the [skin on the hands]."[1]

Upon learning of these results, Pierre Curie applied a radium preparation to his arm for ten hours. For his efforts he ended up with dead flesh inside an open sore. Curie's colleague Henri Becquerel unwittingly joined the self-experimenters when he carried a radium sample in his vest pocket, which branded him with a painful burn. "I love it [radium], but I owe it a grudge,"[2] he exclaimed. Other curious persons subjected themselves to radium burns.

FROM BURNS TO TREATMENTS

If radium could burn or kill skin, perhaps it could destroy tumors. X rays were already being used to treat cancer, since they destroyed fast-growing tumors more readily than healthy cells. Physicians eager for new weapons to try against this deadly scourge added radium to their arsenal, creating the field of radium therapy, or "Curie therapy."

Radioactive sources held an important advantage over x rays. They could be encapsuled and inserted into the body in places where the bulky x-ray apparatus could not reach. These sources could be defined sharply, limiting destruction of healthy tissue. Since patients required only brief exposure times, less expensive radioelements with shorter lives (like mesothorium) could be substituted for radium. Thorium and solid radium decay products (the active deposit) were also used.

Radioactive materials could be applied directly to diseased tissues. It was safer to enclose them in sealed glass tubes. The tubes could be placed on the skin or inside body cavities. For deep-seated cancers, therapists used glass needles and goose quills that could be inserted into the tumor. Though cancer was the highest-priority target for radium therapy, skin problems, lupus, and arthritis were popular candidates. Some physicians and less reputable entrepreneurs tried to cure almost any ailment with radium (Figure 10-1).

Together with cancer, tuberculosis was the era's most feared disease. This illness reached epidemic proportions in the wake of the Industrial Revolution. Population growth strained agricultural resources, inducing many rural workers to migrate to the cities to work in factories. In the United States waves of immigrants from Europe heightened crowded conditions in the cities, allowing the disease to spread readily.

Tuberculosis spared no group or class, striking down the wealthy and well-connected and the poor and powerless alike, often in the prime of youth. Those who could afford the available palliative treatments frequented spas and sanatoriums in hopes of eluding a prolonged and painful death. In 1903 Frederick Soddy suggested treating tuberculosis patients by having them inhale radium or thorium emanation. Unfortunately, this did not prove to be the long-awaited cure, which came only after new antibiotics were developed later in the century.

Soddy's method was adopted to treat other diseases, as people soon realized that radioactive gases offered advantages over the solid materials. A medical facility could fill many glass tubes, bulbs, or needles with radium emanation from a single sample of radium without giving up the precious specimen. By minimizing handling of the radium, they reduced the risk of losing it from breakage or by

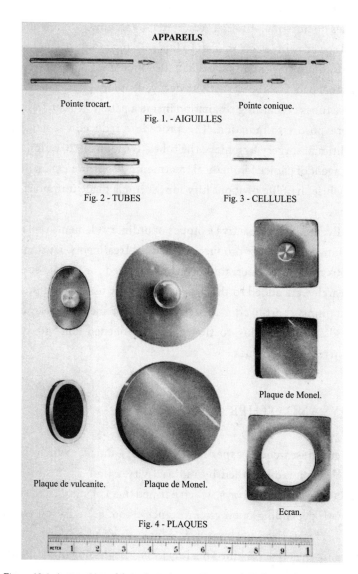

Figure 10-1. Apparatus used for radium therapy. From *Radium* (Brussels: Radium Belge [Union Minière du Haut Katanga], 1925), opp. p. 27.

discarding it accidentally, for instance with medical waste. Many times doctors and physicists resorted to searching through dumps and sewers with electroscopes to find missing radium.

Radium emanation's short half life (about four days) presented another benefit. Since its activity decayed rapidly, physicians could leave tubes containing emanation inside a patient without risking overexposure to radiation. This procedure reduced the need for additional surgery to remove the tubes. By about 1920 radon was the agent of choice for medical treatments. (Thorium emanation's very brief half life of about fifty-four seconds made it impractical to use.)

Eventually, radioactive isotopes of ordinary elements replaced radium and its decay products in medical treatments. During the 1950s new isotopes created in machines used for atomic energy research were added to the list of options. By using small probes containing short-lived radioactive isotopes, medical personnel could target radiation to the exact spot needed and minimize harm to healthy tissues.

RAYS AND OTHER ORGANISMS

Several researchers experimented with animals, which were quickly injured or killed by radioactivity. Microorganisms also reacted to the radium rays, which harmed them or stimulated their growth depending upon circumstances. Though radiation could kill bacteria, it was not appropriate for curing bacterial infections because the rays also killed healthy tissue. Beginning in the 1950s radioactivity's germicidal action was used to sterilize food products, a controversial procedure.[3] Though effective at eliminating bacteria, the energetic rays also cause changes in genetic material and other proteins. Since the long-term effects of ingesting these changed food products are unknown, many people have argued against taking that risk.

After burning his arm with radium rays, Giesel experimented with plants. Leaves exposed to radium lost their healthy green color and turned brown. Apparently, radioactivity could kill plant cells. On the other hand, experimenters showed that small doses of radioactivity stimulated plant growth. Suddenly, waste from radium and uranium factories became valuable for fertilizer manufacturers. Besides triggering growth and early maturation, products such as "radioactive manures," French *engrais radioactifs* (radioactive fertilizer), and German "Radioaktin" reportedly increased crop yield and resistance to weather and other plant hazards.

If radioactivity could benefit plants and other organisms, couldn't it be good for people? Perhaps radioactivity's stimulating effects would improve general health. At first, radiation does increase red blood cell production. This response could improve a person's color, increase energy, and create a sense of well-being. In hopes of curing a variety of ailments, some physicians injected patients with solutions of radium or radon or prescribed these concoctions for patients to drink. With only the fuzziest of rationales and no empirically based standards for treatment or evaluation, radium therapy was an easy target for pseudoscience and charlatans.

MIRACLE CURE?

During the early twentieth century, cancer held an almost mythical status in the psyche of the general public. Like an invisible monster, cancer destroyed its victims while their loved ones and physicians could only stand by helplessly. The promise of a cure for this dreaded disease excited the popular imagination and fueled a public craze over radium.

Radium quickly progressed from a remedy for cancer to an all-purpose magic potion, the miraculous substance promising to cure every ailment. Desperately ill patients and chronic sufferers fixed their hopes on the new wonder drug. Entrepreneurs met the demand with patent medicines and expanded their market with toiletries such as toothpaste and cosmetics, all supposedly containing radium. Many claims were bogus, but a few preparations really did contain radium, to the detriment of their users. Little did people realize that the miracle element could destroy health and cause the very disease they most feared.

RADIOACTIVE SPAS

During the nineteenth and early twentieth centuries health spas were enormously popular. These spas exploited natural water sources reputed for healing properties. Medical professionals routinely prescribed spa treatments for chronically ill patients. Spas were also favored destinations for vacations and for making social, political, and business contacts.

The tradition of healing waters is widespread and very ancient. Many Old World spas date back at least to Roman times. The New Testament mentions Jerusalem's pool Bethesda (John 5). Ireland's countryside is dotted with legendary holy wells, and India's Ganges River is said to have healing power. Native Americans used water sources in the New World for cures. Hundreds of springs and wells reputed to have medicinal properties were scattered around the United States alone.

During the rush early in the twentieth century to find radio-active substances in the environment, researchers tested different waters for radioactivity. Two Italian physicists, Alfonso Sella and Alfredo Pochettino, detected radioactivity in air percolated through water in 1902. J. J. Thomson obtained similar results with tap water the next year. He also found a radioactive gas in well water. Several researchers determined the gas in question was radium emanation.

Radium emanation turned up in spring water and well water all over the world, including water at famous spas like Bath in England, Baden-Baden in Germany, and Hot Springs in Arkansas. A few spa waters also contained radium. Many newcomers to radioactivity searched for radioactive waters.

The search also attracted seasoned researchers. Pierre Curie and Albert Laborde tested waters from nineteen French and Austrian hot springs. Most were radioactive. Julius Elster and Hans Geitel, pioneer investigators of environmental radioactivity, found that sediments from Baden-Baden's springs were radioactive.

Mineral waters had long been touted for their medicinal properties, but physicians had not been able to fully explain their powers. Though some thought the cure was in the mind rather than in the water, spas were a mainstay of nineteenth- and early twentieth-century therapy. Could radioactivity account for some of their legendary healing powers? This idea occurred to scores of scientists and physicians. Though not proven, the cachet of a radioactive cure enhanced the spa mystique. Established spas proudly advertised their waters' radioactivity. St. Joachimsthal, the village near the pitchblende mines that first revealed radium's existence, turned itself into a spa town that offered treatments with radioactive water.

An unusual case occurred in northeastern Oklahoma and southeastern Kansas, where the word *radium* became so closely associated with *cure* that spas advertised "radium water" without pretending that the water contained radium. During the first decades of the twentieth century, health seekers came from all over the United States and even from abroad for treatments with a highly mineralized sulfurous water known locally as "radium water." The term was used to signify the water's healing properties rather than to identify its chemical contents.[4]

Instead of flowing from an ancient source, this radium water burst out of a failed venture to produce oil and gas. In 1903 a newly drilled well in Indian Territory (later northeastern Oklahoma) spouted forth a foul-smelling water that caused paint on houses to peel and turned metal black. This disappointing result transformed into an economic opportunity after analysis of the water showed, in addition to the hydrogen sulfide responsible for the unpleasant effects, many minerals thought to have curative properties.

Enterprising businessmen built bathhouses, creating an area known as "Radium Town" in the city of Claremore. The business spread to other towns in northeastern Oklahoma and southeastern Kansas. Radium water put Claremore on the map long before her favorite son, the comedian Will Rogers, became famous. Rogers extolled the virtues of his adopted hometown's major industry (Figure 10-2).

Though locals believed the water contained no radium, tests in the 1950s revealed its presence. By then the business was beginning to decline. Coming at a time of rising concerns about radium's potential for harm, the ironic discovery that "radium water" actually contained radium was considered a liability and did not help the business.

Development of new medicines as well as social and cultural changes caused the spa movement to wane. New treatment methods plus recognition of the dangers of ionizing radiations brought

Figure 10-2. Radium Town. Photo courtesy of Claremore Daily Progress.

an end to radium's use as a cure-all. Most of the radium baths, tonics, and other fads disappeared; but radiation therapy remained a mainstay of cancer therapy, generally using artificially produced radioactive isotopes instead of radium.

In a few areas, for instance at St. Joachimsthal (now known by its Czech name, Jáchymov) and in the Rocky Mountains, treatments with radioactive waters remain available in the twenty-first century. Naturally occurring radon gas is touted as a healing agent at Bad Gastein in the Austrian Alps. Proponents of such treatments often cite studies which suggest that low-level radiation can benefit health. Such opinions are more common in Europe than in the United States, where the official view is that no level of radiation can be guaranteed safe. The significance of studies on this issue has been widely debated.

DANGERS IN THE LABORATORY

Though radioactivity was hailed for its healing potential, it took a toll on those who worked with it. In addition to causing burns and other skin injuries, radioactivity harmed the eyes, nerves, lungs, liver, bones, and blood marrow, ultimately taking the lives of some researchers. No one understood the hazards at first, since problems caused by radiation often require years to develop. The visible injuries were regarded as minor nuisances and were considered to be part of the cost of doing research.

Scientists handled radioactive preparations with their bare hands and discarded containers in the trash. They spent long hours in closed rooms permeated with radon and carried radioactive samples in their pockets and suitcases. Some staged flame tests with radium, sending radioactive particles around the laboratory or lecture chamber. Lise Meitner and Otto Hahn kept a crate of uranium salts under their work table.

Occasionally a glass vessel containing radium or polonium would explode, endangering the experimenter, scattering precious material, and contaminating the area. Alpha particles were the culprit. These weakened the containers as they bombarded the glass. They also decomposed any moisture in the vessel into a flammable mix of hydrogen and oxygen gases. Over time the pressure inside a sealed container would become dangerous and cause an explosion.

The Curies were especially cavalier about handling radioactive substances and reluctant to admit that radioactivity could cause serious medical problems. Early in their investigations, the Curies suspected their labors were affecting their health. After she had a miscarriage in 1903, Marie Curie feared her work may have compromised her baby's life. During the months after the couple received the Nobel Prize they felt increasingly unwell and were plagued with negative feelings. They endured nerve pain in their fingers and constant fatigue. The birth of their daughter Eve in 1904 did not seem to bring Marie and Pierre the joy they had expected, for they felt old and tired. They rationalized their physical problems away, blaming them on overworking.

During the 1930s Marie Curie became almost blind from retinal damage and cataracts caused by radiation. Laboratory workers also developed illnesses, and one (Sonia Cotelle, née Slobodkine) died after an accident with polonium. Several other former laboratory workers and colleagues met the same end. In 1925 Curie had collaborated on a report for the French Academy of Medicine that recommended safety measures for industry. Safety precautions became part of the routine in Curie's laboratory. Yet, Curie still resisted the thought that radioactivity was causing lasting illnesses in her laboratory and did not act decisively to protect herself. Though she expressed doubts privately, she was reluctant to blame radioactivity for chronic ailments or to make major adjustments to laboratory routines.

In line with her early asceticism, Curie implicitly believed that suffering was the price to pay for furthering the advance of science. At some level Marie Curie was following in the path of the saints and martyrs of her childhood religious education, but with a different ultimate goal. Her childhood models pursued detachment from worldly goals and a single-minded purity of heart leading to holiness. Curie's goal was a detachment from ordinary concerns and a single-minded pursuit of pure science leading to scientific truth. Both paths required suffering.

If researchers did not take precautions for their own safety, they might do so for the sake of their experiments. Radioactive gases could quickly contaminate an entire laboratory with their decay products, making it impossible to take meaningful measurements.

For instance, a researcher (May Leslie) noted in 1909 that the room where Marie Curie worked was quite radioactive.[5] Arthur S. Eve did measurements at home in order to avoid the omnipresent radioactivity in Rutherford's McGill laboratory. Meitner and Hahn were happy to move from their radioactive woodshop to the new Kaiser Wilhelm Institute, and they took strict measures there to prevent contamination.

Even after they realized that stray radioactivity could ruin experiments, many researchers continued to be careless. The Paris Radium Institute became contaminated, as did a room in the Vienna Institute where Otto Hönigschmid was accustomed to shake his radium solutions by hand. Eventually, researchers in Vienna's Radium Institute had to make their measurements in a different building. Notebooks of Becquerel, Marie Curie, and Sir William Crookes were still radioactive a century later.

The list of scientists injured or killed by radioactivity is extensive. Stefan Meyer gave up playing the bass viol because radium had damaged his fingers. Hahn's fingers were also harmed by radioactivity. The physicist Marietta Blau suffered radiation injury to her hands and eyes from working in the Vienna laboratory. Giesel (whose breath had contained enough radon to discharge an electroscope), Hönigschmid, Egon von Schweidler, and André Debierne all died of lung cancer. Georg von Hevesy also succumbed to cancer. Marie Curie and her daughter Irène died from blood diseases caused by radiation, while Frédéric Joliot developed fatal liver cirrhosis from working with polonium. Sonia Cotelle's misfortune has already been mentioned. Other victims include Catherine Chamié and Marguerite Perey of the Paris laboratory and numerous individuals involved with uranium mining, the radium industry, and radium therapy. This list is not exhaustive, but shows the hazards faced by early researchers.[6]

The Paris and Vienna laboratories experienced a disproportionate share of radiation-related deaths. One reason may be that these institutions were heavily involved with separating and purifying radioactive substances and other chemical operations that required workers to manipulate radioactive compounds. The chemical work, as opposed to physical investigations with purchased radioactive substances, created multiple opportunities for workers to be exposed to and to ingest radioactive substances. Laboratories in Paris and Vienna were well placed for chemical research because radium was more accessible to them than it was to other institutions. The Curie laboratory also did extensive research with polonium, now known to be especially dangerous in the human body.

Though warning signs were present for anyone who wished to heed them, during the first two decades of the twentieth century most people ignored or minimized these signs. As early as 1905 a French radium therapist succumbed to radioactivity. Why did this death, and the multitudes of problems and suspicious ailments afterwards not provoke a concerted safety effort?

First, since radiation's most damaging effects take time to appear, it was easy to assume that problems would be minor and temporary. A death could be considered an anomaly caused by the unwise actions of an individual, rather than a systemic problem. Second, the newness of the field, its dispersion among private and public entities in different countries, and the freedom to study and use radioactive substances by anyone who could get them meant that no generally accepted standards or regulations were in place to limit dangers and abuses. Third, publications often produced contradictory results on whether radioactivity harmed or improved health. This happened both because of variations in constitutional resistance to radiation damage and because controlled, standardized studies were not available. Finally, human hopes and ambitions colluded to minimize the perceived dangers. A certain stoicism was also in the mix, most strikingly with Marie Curie but common among radioactivity's pioneers. Interviewed in the 1960s, Hahn remarked that "the danger [of working with radioactive substances] is exaggerated nowadays....I am always rather suspicious of people who make a great fuss."[7]

For scientists, the excitement of a new field which brought surprises almost monthly was not to be dampened by vague fears. Hesitation might hamper the speedy work crucial for career advancement. Worse, it could block a discovery or delay development of urgent medical cures. Medical practitioners as well as many industrialists (some of whom were motivated by a relative's illness or death from cancer) were anxious to help their desperate patients. Radioactivity's promise as a healer blinded the new field's pioneers to its dark side. Scientists, physicians, and industrialists alike would find it hard to admit they might have unwittingly caused harm. Financial interests also predisposed some to believe in the safety of their products.

A few radiation effects, like skin problems and other suspect symptoms, were so common and well-known that most individuals and institutions adopted rudimentary safety measures during the second decade of the twentieth century. Though inadequate by modern standards, lead shielding was used widely. Some companies limited their employees' contact with radioactive substances and improved the ventilation in work areas. Not until the 1920s and the 1930s, after the International Radium Standards Committee developed measurement standards and more data on radiation effects became available, did professions adopt more rigorous guidelines for radiation safety.

New Industries

Nothing is an unmixed blessing.

—*Horace, Odes*

EARLY INDUSTRY

At first, most radium was destined for scientific research. Since this element was quite rare, only those who were well connected or well financed could obtain it. The German chemist Friedrich Giesel and the Curies supplied radium to their colleagues. Giesel had significantly reduced the time needed to prepare radium by using bromide compounds instead of the chlorides used by the Curies. Using this technique, he began marketing radium in 1902 through his employer Buchler and Company, a quinine manufacturer in Brunswick, Germany. Giesel also prepared an impure polonium. Another German firm, owned by Eugen de Haën and located near Hanover, had supplied Giesel with materials and began selling radium in 1899.

In 1899 Pierre Curie and André Debierne arranged for industrial processing of radioactive materials with the Société Centrale de Produits Chimiques (Central Society of Chemical Products), with limited resources. The industrialist Armet de Lisle then offered the Curies funding and space in his factory near Paris, enabling them to produce larger quantities of radium. In order to meet the growing demand from physicians, his factory began commercial production in 1904. A Hamburg firm manufactured polonium according to Willy Marckwald's methods.

This promising situation deteriorated when the Austrian government began raising the price of pitchblende as it prepared to process its own ores. Radium's price quickly skyrocketed, making it unaffordable for many researchers. The German factories could no longer produce polonium. Wishing to preserve their ore, which was being frittered away for nonscientific purposes, and perhaps hoping to corner the market, Austria banned the export of pitchblende late in 1903. (An exception was made for the Curies.)

Frederick Soddy reacted with consternation and jumped to conclusions. "Do you know," he wrote to Ernest Rutherford, "I have a shrewd suspicion Curie has ... secured the monopoly of the Joachimsthal mine, i.e. the only practicable source of radium, damn him. The Austrian Government have [sic] closed down on supplies of the residues.... The German people write me that the residue is not to be had."[1]

Though Austrian scientists could get native pitchblende, they needed a way to extract the radium while the government operation was being planned. The Austrian Academy of Sciences made arrangements to refine 10,000 kilograms of pitchblende residues at a lighting factory owned by Carl Auer von Welsbach. This famous chemist had founded a thriving business manufacturing devices (incandescent mantles made with thorium) designed to increase the effectiveness of gas lights. Around 1907 the Austrian government completed a radium factory at St. Joachimsthal. Investors in the United States also developed businesses to utilize native ores.

Frustrated with the unavailability of pitchblende, the German Association for Applied Physical Chemistry recommended in 1907 the creation of a radioactivity institute in Austria. Such an institute would allow Austria to control its own ores while permitting foreign scientists to pursue research. This vision was implemented in 1910 upon the initiative of a wealthy donor.

Mesothorium (later known as mesothorium I) also became industrially important, giving German manufacturers an alternative to the essentially unavailable pitchblende. A by-product of thorium extraction, mesothorium behaved like radium chemically but had a much shorter half life. Since mesothorium was less expensive to produce than radium, manufacturers often used it as a radium substitute. Its activity did not last nearly as long as radium's, so that pure mesothorium preparations would weaken noticeably after a few years. Commercial mesothorium containing some radium performed better. A few years later scientists realized that mesothorium I was an isotope of radium, chemically identical to radium but with a different atomic weight (Ra^{228}; the first radium discovered was the isotope Ra^{226}).

Mesothorium was first marketed in Germany, earning it the sometimes derisive nickname "German radium." Otto Hahn had advised the world's largest thorium producer, the Berlin firm owned by chemist Oskar Knöfler, to produce mesothorium to quench the ballooning medical demand for radium. Knöfler's company and Auer von Welsbach's Incandescent Light Company in Austria also sold mesothorium to manufacturers of luminous paints. Starting in 1911, the French-Brazilian Industrial Mining Society produced mesothorium with the guidance of Marie Curie and Debierne.[2]

SOARING DEMAND AND NEW INSTITUTIONS

Originally discovered in 1789, uranium's main industrial use had been to color ceramic glazes and glass. Fiesta ware, a bright orange pottery, and vaseline glass, a yellow glass with green fluorescence, were popular twentieth-century examples of such products. From about 1910 the drive for cures, especially cancer cures, created a tremendous need for radium and its parent, uranium. Suddenly, uranium was highly prized.

Because Austrian ore was unavailable, entrepreneurs turned to other sources. They mined uranium ores in places as diverse as Cornwall in southwest England, Portugal, Norway, Turkistan, Germany, Madagascar, Japan, Ceylon, and Australia. Around 1910 the United States began marketing carnotite, a uranium mineral first found in Colorado and identified in 1899.

Early attempts to prepare radium commercially failed, leading Americans to ship most of their ore to Europe for processing. Since carnotite ore beds in Utah and Colorado were extensive and relatively easy to extract, the United States soon became the world's largest uranium supplier. "The uranium deposits of Colorado and Utah are being rapidly depleted for foreign exploitation," warned the Bureau of Mines. "It would seem to be almost a patriotic duty to develop an industry that will retain the radium in America."[3] Realizing that their ores were subsidizing foreign industry, the U. S. government tried unsuccessfully in 1914 to nationalize all new mines. By then several domestic companies were producing radium.

As demand continued to rise, radium became extremely costly, with its price per gram increasing from about $2,500 in 1904 to $120,000 in 1913. Private donors, scientists, and governments intervened to support research by founding specialized institutes for radioactivity. The field's undercurrent of national rivalry took visible form in these buildings and institutional structures.

The first, the Institut für Radiumforschung (Institute for Radium Research) in Vienna, opened in 1910, thanks to the beneficence and foresight of a lawyer and industrialist, Karl Kupelweiser. Concerned that Austria's pitchblende, a "treasure" given to her by nature, was being exported while Austrian physicists could not afford radium extracted from this ore, Kupelweiser donated a large sum (500,000 Austrian kronen) to the Viennese Academy of Sciences to construct a building for physical research on radium.[4]

In Germany, plans were made for a series of scientific institutes to be supported by industry and government. The first, the Kaiser Wilhelm Institute for Chemistry, opened in 1912. Hahn headed a small radioactivity department in the institute. In 1913 Lise Meitner received an official position matching Hahn's, and the radioactivity section became known informally as the Hahn-Meitner laboratory.

The Polish Academy of Sciences created a laboratory for radioactivity in Warsaw. The planners had hoped to entice their famous native daughter back to Poland by offering her the directorship of the laboratory. Marie Curie was torn between her responsibilities to her beloved homeland and to her French laboratory. She finally agreed to accept the directorship, but to assign control of the institute to her Polish assistants Jan Danysz and Ludwig Wertenstein. Curie traveled to Warsaw for the building's formal opening in 1913.

Pierre Curie's dream of a modern laboratory finally materialized when the Institut du Radium (Radium Institute) opened in 1914. Funded jointly by the University of Paris (the Sorbonne) and the Pasteur Institute, it consisted of two independent laboratories: one for radioactivity, to be directed by Marie Curie, and one for biological research and medical therapy, to be run by the physician Claude Regaud. Not one to assume that others would be able to design the laboratory she and Pierre had so long desired, Marie Curie had interpolated herself into the architect's planning process. She took a special interest in creating a garden between the two institute buildings, a monument to her lifelong love of nature.

A few weeks later, before Marie Curie could even move her equipment into the new building, Germany invaded France. Curie spent much of the next few years supporting the French army by requisitioning, designing, installing, and operating both fixed and mobile x-ray stations for diagnosing the wounded.

In addition to the research laboratories, radium institutes were formed to garner the scarce element for medical treatments. Britain and the United States opened medically oriented institutes before World War I. After the war, similar radium treatment centers sprang up in places as far flung as Montreal and St. Petersburg.

The U.S. industry suffered during the war, when money that European customers normally spent on radium was allotted for military needs. Business improved temporarily when the U.S. military forces began ordering luminous paint made with radium. Manufacturers ramped up production, only to find themselves with a surplus which domestic demand could not meet. After the devastating world war, foreign purchases could not make up the difference.

A challenge came from a new supplier in 1922. This could not have happened at a worse time for the radium manufacturers. Rich uranium ores had been discovered in central Africa in 1913 and 1915, in what was then called the Belgian Congo. The Belgian Mining Society's factory, in Oolen near Antwerp,

Figure 11-1. Crystallization vats for purifying radium, early 1920s. From *Radium* (Brussels: Radium Belge [Union Minière du Haut Katanga], 1925), opp. p. 2.

began processing this pitchblende ore in 1922 (Figure 11-1). The Belgian factory could produce much more radium than American producers, who were limited by their lower-grade ores. Burdened with unprofitable stockpiles of radioactive materials as the nation entered the Great Depression, American companies could not compete with the newcomer. By 1923 Belgian radium had saturated the market, causing the price to fall to about $70,000 per gram and edging out the world's other radium industries.

A competitor arrived in 1932, when producers began extracting radium from high-quality Canadian pitchblende. The rival Canadian and Belgian businesses decided to cooperate. They divided the radium market to avoid competition, and set radium's price at $40,000 per gram in 1938.

PAINT THAT GLOWED IN THE DARK

Next after medical applications, radium's most important industrial use was for luminous paints made with phosphors. These
paints were developed in the nineteenth century, when entrepreneurs hoped to illuminate the night with something more convenient and efficient than candles, arc lamps, or gas light. Phosphors
are materials that will glow in the dark after they have been illumi-
nated (or activated in some other way). Paint containing phosphors
could not supply the intensity required for large applications, but
it became popular for clock and watch dials, compasses, and signs,
as well as for decorative and novelty objects. Since phosphors normally did not emit heat, they were economical and efficient. But
their "cold light" had a fatal drawback: it faded with time. To provide light throughout the night, phosphorescent paint would need
periodic recharging. Using another light source to do this would
defeat the paint's purpose.

Radioactivity promised a solution. In 1897 Ivan I. Borgman,
professor of physics at the University of St. Petersburg in Russia,
noticed that radioactive substances made phosphors glow. Several
years later (1902), Giesel found that screens made of a crystalline
zinc sulfide known as "Sidot's blende" glowed very intensely when
struck by alpha rays. In 1903 Sir William Crookes, and independently
Julius Elster and Hans Geitel, discovered that the screens' phosphorescent glow was made up of discrete bursts of light (scintillations).

Most researchers saw radioactivity's effects on phosphorescent screens as clues to the nature of the rays. A few realized these
effects also presented a way to prolong phosphorescence. If radium
were added to a phosphorescent paint, the product might glow
indefinitely, since radium would continue supplying energy to it
for thousands of years.

Phosphorescent paints containing radium will not actually last this long, because the powerful radiations eventually destroy the phosphors. However, adept manufacturers produced paints that could serve for many years. Giesel's firm began marketing luminous paint containing radium in 1906. Others followed, sometimes using mesothorium (later known as mesothorium I), radiothorium, and mixtures of these radioelements with radium as less expensive alternatives.

During World War I the phosphorescent paint industry thrived as military forces on both sides of the conflict demanded luminescent switches, sights, watches, and instrument dials. Luminous paints were also used on signs, markers, buoys, and even soldiers' collars. With these aids troops could maneuver without using bright light sources that would reveal their locations. In the United States, the Radium Luminous Materials Corporation in New Jersey (later the U.S. Radium Corporation) was the largest supplier of luminescent dials.

Clocks and watches with radium-laced hands and numbers became popular among civilians as well, driving the industry after the war. Approximately 110 firms purchased paint from the U.S. Radium Corporation in 1925. The clock manufacturing centers in Connecticut, New York, and Illinois were major clients. Dial painting was a coveted factory job, as it required no heavy labor and did not involve high temperatures, irritating fumes, or dangerous machinery. Women, whom employers considered most suitable for careful, detailed work, usually filled these posts.

A NEW POISON

During the 1920s many workers in these plants became ill with mysterious maladies. Fatigue and general malaise often progressed to death of jaw tissue and unusual cancers. Some dial painters and physicians suspected that these illnesses were related to the painters' occupation. Employers denied any such connection.

As more dial painters were stricken with chronic illnesses, evidence mounted for an occupational link. A likely candidate was the technique of forming the brush with the mouth and lips. Each time a painter put her brush between her lips, she swallowed a little radioactive paint. By allowing the paint to get inside her body, she greatly increased her exposure to radioactivity. The painters' bones readily assimilated the radium in the paint, since this element resembles calcium chemically and calcium forms part of bone structure. The embedded radium emitted alpha particles, which gradually destroyed the unfortunate workers' bodies from within. Many died gruesome and painful deaths, while others were disfigured or incapacitated.

Radium killed other Americans who worked with it, including both the chemist (Edwin Lemen) and the founder (Sabin von Sochocky) of U.S. Radium Corporation. Lemen's widow and some of the dial painters sued the company. The media pounced on these cases and aroused public sympathy particularly for the suffering women, who were often struck down in the prime of youth. In 1928, after lengthy legal battles, U.S. Radium agreed to pay for injuries and deaths of some of its workers. Other firms fought lawsuits brought against them or settled out of court.

The practice of pointing the brush with the lips ended in the mid-1920s, abating the stream of deaths. Other sources of radium poisoning were not eradicated so readily. Workers continued to inhale the radon and paint dust which pervaded the dial painting studios. Radioactive dust and debris contaminated their hair and clothing. Manufacturers often ignored recommendations to clean up the work spaces. In addition to radiation illnesses and direct tissue destruction, cancer began to appear among dial painters and other exposed workers. It would be many decades before adequate safeguards for factory workers were enforced.

During the 1920s problems also emerged with patent medicines and other nostrums supposedly containing radium. A study by the U.S. Department of Agriculture's Bureau of Chemistry published in 1922 revealed that most of the "radioactive" mineral waters tested contained no radium, and therefore had no therapeutic value derived from it. A 1926 report followed up with an analysis of patent medicines, cosmetics, and other preparations. Only five percent of the products tested contained "therapeutic" amounts of radium. Unwitting consumers were being duped by misguided or unscrupulous entrepreneurs. False advertising was rife in these industries, where manufacturers could save money by selling "radium" cures minus the radium.

The swindled customers were fortunate. By the late 1920s the dangers of ingested radium were becoming clear. Though earlier studies had pointed to health hazards associated with radioactivity, these reports inspired only minimal changes in procedures. The widely publicized case of the dial painters was impossible to ignore. Another well-known instance of radium poisoning involved a tonic called "Radiothor" that contained both radium and mesothorium. Radiothor was marketed during the 1920s and became popular among the upper classes. This product became suspect after it caused the illness and subsequent death of a prominent American socialite in 1932. That year the American Medical Association withdrew its approval of radium for internal use.

The newfound respect for radioactivity's dangers did not immediately transfer to everyday life, where radioactivity was still seen as beneficial. Rare and mysterious, radium's double-edged power to both help and harm enhanced its allure. However, by the 1930s, radioactivity's long-term effects were starting to show up in the premature deaths of scientists who worked with radioactive materials. Uranium miners also had unusually high rates of lung cancer. Still, the overall public image of radioactivity remained positive.

FISSION, BOMBS, AND THE URANIUM RUSH

In 1938 chemists Otto Hahn and Fritz Strassmann announced from Berlin that, in certain experiments, uranium produced a lighter-weight element, barium. They had not expected this result, but were sure that they had correctly identified barium. The physicist member of their team, Lise Meitner, had recently fled to Sweden to escape the Nazis. Meitner's nephew Otto Frisch, also a physicist, happened to be visiting Meitner when she received the news. Meitner and Frisch were able to explain what had happened. The uranium nucleus had split into two pieces, freeing large amounts of energy.

Frisch chose the name "fission" for this reaction, borrowing the biologists' word for cell division. Several scientists immediately recognized that, under the right circumstances, fission could become a self-sustaining reaction, one that would continue indefinitely. By building a machine, or reactor, to control this process, society could gain a new, long-lasting energy source for industry and for generating electrical power. On the other hand, by developing a way to provoke an uncontrolled fission reaction, society could add a new method of mass destruction to its arsenal, a nuclear-powered bomb.

The obvious geopolitical implications of such a bomb led Germany, France, Britain, and the United States to embark on secret programs to harness nuclear energy. Germany was already heading towards a war of expansion and dominion. The other nations assumed that Germany, with its top-notch physicists and educational system, held an advantage and would use a nuclear bomb to further its nationalistic ends. Due to a mixture of bureaucratic obstacles and scientific missteps, Germany did not succeed. The United States' crash program, called the Manhattan Project, came to fruition in 1945. After testing a uranium bomb in New Mexico, the project produced bombs that exploded over Japan in August, hastening the end of the war and astounding the world.

The nuclear bomb shattered any remaining notions that physics could remain aloof from politics. The explosions caused unprecedented destruction and death in Japan and succeeded in evoking widespread fear for the world's future. This act was widely seen as ushering in a new age, the atomic era. Debates over the rationale for using the bomb in Japan and the morality of nuclear weapons in general continue to this day.

After the fateful explosions in 1945, radium's image developed openly sinister overtones. Authorities gradually put more stringent safety regulations in place for handling and storing radioactive substances. Radioactivity's presence in the environment, as in fallout from bomb tests and radon in homes, became topics of public concern.

During the postwar period, the United States began stockpiling uranium bombs for protection against the perceived menace presented by a Soviet Union armed with nuclear weapons. The 1950s saw the height of the Cold War, a period of mutual distrust marked by spying, secrecy, bomb shelter drills, and fanatical searches for traitors. The federal government promoted uranium prospecting and allowed uranium to be sold only to the new federal Atomic Energy Commission (AEC).

Eager adventurers streamed onto the Colorado Plateau hoping to strike it rich. The population of remote areas swelled. New businesses sprang up to accommodate the visitors and cash in on the prospecting fever. As geologists and others pondered ways to locate uranium ore, the oil industry entered the scene.

RADIOACTIVITY AND THE OIL INDUSTRY

Early in the twentieth century researchers had learned that petroleum deposits were slightly radioactive. At the Freiburg Physics Institute in Germany, Heinrich Rausch von Traubenberg noticed that petroleum readily absorbed what he called "radioactive air." Following up on this observation, the institute's director, Franz Himstedt, checked petroleum samples and found they were radioactive. Across the Atlantic, Eli F. Burton at the University of Toronto detected a radioactive gas in petroleum. Researchers later identified the gas as radium emanation. Others observed that helium often accompanied petroleum deposits. Both radium emanation and helium were products of radium's decay. Apparently petroleum contained traces of radium.

Oil's radioactivity was no more than a curiosity until the 1950s, when finding uranium became a national priority in the United States. If petroleum could harbor radium, perhaps uranium would be found near oil deposits. Prospectors and geologists combed the "oil patch" with Geiger counters to search for radioactive minerals. They analyzed rocks and geologic formations that contained uranium and proposed physical and chemical reactions that might concentrate radioactive elements in petroleum deposits. For a time it looked like petroleum might be a marker for uranium deposits.

These hopes were short-lived. Searching for uranium based on the location of petroleum did not prove efficient. Though petroleum-bearing rocks and other hydrocarbon deposits often contain a little radium and uranium, uranium deposits are not usually located near petroleum reservoirs. A geochemical process is responsible for the small amounts of uranium and radium found in petroleum bearing-materials.

The reverse procedure, using radioactivity to search for oil, had limited value. Since radioactive materials are widely dispersed in the earth's crust, finding radioactivity does not necessarily mean oil is nearby. Surveys of radioactivity and maps of radon occurrences have been most effective when combined with other prospecting methods.

Petroleum did not prove to be an economically viable source of radioelements. Instead, radioactive materials recovered during petroleum processing created additional problems and costs for the oil industry. Not being a profitable source, they were treated as hazardous waste.

Though a few lucky and persistent entrepreneurs succeeded in finding marketable uranium during the 1950s, the boom was short-lived. After the nuclear bombs were built and stockpiled, uranium saturated the market, and the bubble burst. The AEC changed its focus to nuclear power, but health and safety concerns collapsed this industry in the 1980s. Energy shortages in the early twenty-first century offered an opening for nuclear power. For the industry to succeed, it would have to find ways to minimize the drawbacks of cost, security hazards, and disposal problems.

PART THREE

BEYOND THE STORY

Human affairs do not become intelligible until they are seen as a whole.
—Attributed to Arnold Toynbee

How does radioactivity's history fit into broader patterns associated with scientific activity and with human endeavor in general? What can we learn by looking back at this extended episode?

Numerous factors affected radioactivity's history. Some ingredients were part of the science of a particular era and place, while others can be traced back to the dawn of history. Many continue to influence scientific activity.

Radioactivity was viewed through and influenced by common conceptual patterns used to understand the physical world. The ways questions about the natural world are framed and the kinds of answers considered appropriate have changed historically, but certain features have persisted.

Going beyond science per se, radioactivity awakened longings and strivings of the human spirit that have inspired human activities for as long as records exist. If we cannot conclude that such tendencies are coded into the human brain, they at least have been part *of the human heritage for a very long time. This book's final section describes influences that guided radioactivity's development and explores how perennial questions, patterns, and deep human desires shaped the story.*

Radioactivity's Prime Movers

... it is the genius of man, and not the perfection of appliances, that breaks new ground in the great territory of the unknown.

—*H. J. W. Dam*, 1896

Our society, in which reigns an eager desire for riches and luxury, does not understand the value of science.

—*Marie Curie*, 1923

What created the new field and moved it forward? How was it sustained and transformed? Many forces and factors affected the growth and development of radioactivity as a science. Intellectual, personal, professional, social, cultural, technological, political, and economic influences propelled, hindered, and transformed the new science.

TECHNOLOGY, RESOURCES, AND PROFESSIONAL CHANGES

Radioactivity's discovery followed upon advances in vacuum and electrical technology that enabled experimenters to study electrical discharges in high vacuum. Electrical methods and instruments, especially the electroscope, were crucial for radioactivity's rapid advance. Other new instruments and measurement methods, as well as apparatus made by skilled laboratory technicians, facilitated the field's development.

Researchers needed financial resources to procure radioactive materials, and radium's scarcity restricted its use. Changes within traditional disciplines, new institutions, and new avenues for publication helped to promote the burgeoning new research area. These factors have been discussed in parts I and II.

INDIVIDUALS

The field of radioactivity was created by individuals. Their curiosity, persistence, initiative, and creativity moved the new science forward and brought it to maturity within a remarkably short time. The enthusiasm and boundless energy of Ernest Rutherford, the sharp intellect and initiative of Frederick Soddy, Pierre Curie's detached rigor, and the brilliance and perseverance of Marie Curie identified, analyzed, constructed, and honed the new science. Though they encouraged collaboration and a team spirit, individuals like Curie, Rutherford, and Stefan Meyer continued to dominate the field's progress into the 1920s and later.

These pioneers loved their work. Radioactivity is "a splendid subject to work on," effused Rutherford to Marie Curie.[1] Curie's passion led her to regale colleagues with the details of radioactive separations, oblivious to their boredom. Julius Elster and Hans Geitel's pleasure in lives of scientific pursuit, Friedrich Giesel's persistent experimenting in his spare time, William H. Bragg making a career change because he found certain questions irresistible, the Curies' delight in the beauty of luminescent radium—these are just a sampling of testaments to the researchers' fascination with their vocation. The new science could not have thrived without the motivated, engaged leadership of individual scientists.

RESEARCH GROUPS

Most researchers did not work alone, but were part of a community. The Curies belonged to a tightly knit group of intellectuals engaged not only in science but also in mathematics, literature, and philosophy. After Pierre Curie's death, Marie Curie accepted his academic post. Assisted by André Debierne, she ran a laboratory which attracted students and advanced researchers from developed countries across the globe.

Rutherford's leadership and enthusiasm created a community wherever he worked. His students came from far and wide and made many significant contributions. Soddy guided students at the University of Glasgow for several years. His interests eventually turned from radioactivity to economics and social questions.

The congenial and kind Stefan Meyer built a large and loyal following. He created a collegial atmosphere which encouraged camaraderie and cooperation. "The atmosphere at the institute was most pleasant," explained Elizabeth Rona. "We were all members of one family."[2]

In these laboratories scientists built lifelong friendships and professional contacts. The female researchers in particular often supported one another, since they were a minority in places only recently opened to them. As legal barriers to women's education crumbled, increasing numbers of women pursued advanced degrees in the early twentieth century. Many found their way to the radioactivity research centers.

SCIENTIFIC IDEALS AND CULTURE

Science at the turn of the twentieth century was widely seen as objective and impartial, a noble exception in a world subject to the whims of politics and subjectivism. The ideal scientist sought truth without regard to personal benefit or political considerations. To describe someone as "disinterested" was perhaps Marie Curie's highest compliment. The Curies consciously followed this standard when they refused to profit from radium by taking out a patent. Their industrial and fund-raising activities were designed to support radioactivity research, rather than for personal gain. In the same spirit, laboratory directors opened their doors to aspiring scientists from a variety of nations and social backgrounds. Ideal science was an international venture.

This vision of an objective, detached science commingled with broader cultural movements towards fairness and equality, creating a more favorable climate for less traditional students, especially women, to enter scientific careers. Gradually, European and American universities opened their doors to women.

The socialist and positivist notions that inspired young Maria Skłodowska and brought her from her native Poland to the more progressive Paris spread within Europe. In cosmopolitan Vienna, a culturally rich city where women had been admitted to the University of Vienna since 1897, a vigorous socialist movement developed. After briefly seizing power nationwide in 1919, the Social Democratic Party continued to control Vienna's government until 1934. The socialists' program for gender equality, including equal educational and career opportunities, influenced Viennese culture, producing a hopeful environment for women with scientific ambitions. Prominent academics like the positivist philosopher Ernst Mach and physicists Ludwig

Boltzmann and Franz Exner favored higher education and careers for women.

MENTORS AND MODELS

With Marie Curie's name virtually a household word, scientific careers became easier for young women to envision. In a few fields and universities women had found willing, even encouraging mentors. Botany and astronomy had been receptive to women since the late nineteenth century, but physics, with the exception of luminescence and spectroscopy, was staunchly male territory.

Now radioactivity, a new area with fluid professional boundaries, was an option. Beginning around 1900, women entered laboratories in increasing numbers. Meyer's popular 1916 text on radioactivity listed twenty-six women with publications in radioactivity; the second edition of 1927 listed seventy-nine. He dispensed with the customary courtesy of designating women by titles like "Miss" or "Madame," "since [in 1916] it no longer is so exceptional for a woman to work scientifically."[3]

The women came from Britain, France, central Europe, and North America. The top choices for women and men alike were the Curie laboratory in Paris, Meyer's laboratory in Vienna, and Rutherford's laboratory in Manchester.

The careers of aspiring scholars have always depended on their mentors. In the early twentieth century professors could refuse to allow women to take classes or to work in their laboratories. Even after universities formally opened their doors to women, a reluctant professor could make difficulties. In an era where sex discrimination was not only taken for granted, but did not even have a name, finding sympathetic professors was critical.

These helpful advocates were scattered in various places, but luckily included the directors of the main centers for radioactivity. Marie Curie accepted numerous female researchers to her Paris laboratory, where they were kindly shepherded by her colleague André Debierne. Originally recruited to the field by Pierre Curie, the unassuming Debierne was an excellent chemist who, with Pierre Curie, had narrowly missed the discovery of atomic transmutation. After Curie's death, he penned prescient speculations on the cause and mechanism of radioactivity. When Debierne retired, several women took on administrative duties. One of them, Marie's daughter Irène, had a remarkable career crowned by the Nobel Prize for Physics.

Rutherford mentored his first female student, Harriet Brooks, at McGill University in Montreal. Despite harboring implicit cultural assumptions of his time, the forthright and enthusiastic Rutherford supported female researchers. Many women made research contributions in his laboratories at McGill and Manchester.

The record for sheer numbers of female researchers in radioactivity may belong to Stefan Meyer. Students and colleagues praised Meyer's warm, kind character, his friendliness, and his desire to help fledgling researchers with their careers. During his long tenure dozens of women passed through the Vienna Institute for Radium Research. Some stayed only briefly, to fulfill a requirement for their degree and license to teach; but others made a career in the field. Later, Meyers' daughter Agathe Meyer-Rosenquist reportedly worked with her father on some problems in nuclear structure.[4]

The personal beliefs of Meyer and his elder colleagues Franz Exner and Ludwig Boltzmann together with a favorable political and cultural environment ensured women a welcoming and supportive venue at the Vienna Radium Institute. This happy situation ended after the Nazis annexed Austria in 1938. All Austrians automatically became subject to Germany's nefarious racial laws. Meyer, who was partly Jewish in ancestry, forestalled his inevitable expulsion by resigning. Others who were endangered or who refused to cooperate with the Nazis resigned or were dismissed. Women who remained had to deal with a regime that was philosophically opposed to their presence; the Nazis did not favor women working outside of the home. Meyer's student Bertha Karlik nevertheless took over his administrative work, and was named director after the war's end in 1945.

Women worked with other mentors, including Soddy in Glasgow and Oxford; J. J. Thomson in Cambridge; Willy Marckwald, Otto Hahn, and Lise Meitner in Berlin; Otto Hönigschmid in Vienna and Prague; Georg von Hevesy in Copenhagen; and Kasimir Fajans in Karlsruhe. The independent, iconoclastic Soddy did not have many students; but he believed in equality for women. Soddy coauthored research papers with several women, including his wife Winifred, who was active in the suffragette movement.[5]

One would expect Marie Curie, who had struggled to make her place in an overwhelmingly male profession, to accept women into her laboratory. For Rutherford and Meyer, a healthy self-confidence insulated them from feeling threatened by competent female colleagues. In general, the newness of radioactivity buffered some of the prejudices common in academia. Rather than being confined by tradition, in this new field the leaders were creating the traditions.

Though she was well-known by the general public, Marie Curie was not the only important female role model in radioactivity. Austrian physicist Lise Meitner was exactly eleven years Curie's junior. She began work on beta rays in Berlin immediately after earning her Ph.D. in 1906, the start of a fruitful and lifelong career in radioactivity and later, nuclear physics. A number of women worked in the Berlin physics and chemistry institutes between the two world wars, including Clara Lieber, a chemist from the United States who took part in experiments which led to the discovery of nuclear fission.

Before scientific mentors even entered the picture, family support was crucial for future scientists. Parents guided, encouraged, and often financed their offsprings' aspirations. Curie's parents were educators who expected their daughters and son to excel alike. Meitner was raised in an educated middle-class Viennese family that backed her financially while she pursued her career goals. Having been a teacher, Rutherford's mother was well-disposed towards his education. Without support from their families, most students could not have pursued research careers.

AGE, ATTITUDES, AND AMBITION

Radioactivity was an exciting subject for young researchers and others who hoped to make their mark in a new area, perhaps even to have their name memorialized by discovering a new element. The field promised opportunities for original work, even major discoveries, without the technical and professional baggage of an established field. Students flocked to Paris, Vienna, and Manchester. Others went to smaller centers like Berlin, Munich, Glasgow, and Warsaw. Some came to launch or enhance scientific careers, others to satisfy a degree requirement in an agreeable environment.

Radioactivity's cross-disciplinary nature stimulated creative thinking in the way that new ideas, art, and religions often arise at the crossroads of cultures. As a new area which straddled traditional boundaries, radioactivity disproportionately attracted the young. Having fewer preconceived ideas and routines was an advantage when researchers continually improvised ideas, methods, and materials.

Not all major players were young. Mental flexibility and openness to new ideas were the most crucial qualities. For instance, William H. Bragg began research in middle age. His lack of experience was an advantage, because he was less invested in particular theories and models than many of his contemporaries.

Status and success were important to most researchers. Among the pathbreakers, Rutherford did not easily conceal his ambition, and Soddy and Marie Curie were keen to gain professional recognition. Competition pushed researchers to work harder and faster, both for their teams and for individual credit. The competition took on a nationalistic spirit, with the British, French, and Austrian groups vying to be first and best.

NATIONALISM

During the nineteenth century nationalism became a potent force in Europe, driven in part by increasing competitiveness for trade, industry, and resources. This factor drove much of the twentieth century's history, and radioactivity did not escape its reach. The changing political climate increased the underlying tension between national interests and the ideal of a detached, international science.

The popular press interpreted events in the light of national competition, regardless of the motives of individual scientists. Becquerel's and the Curies' discoveries became kudos for France, Rutherford's for Canada and then Britain proper, Meyer and

Schweidler's for Austria. French, British, Austrian, and German researchers scrambled for priority and primacy. The title of a physicist's 1921 essay, "The Part Played by Different Countries in the Development of the Science of Radioactivity," shows that this kind of informal tally sheet was considered an appropriate way to classify radioactivity's accomplishments.[6]

Marie Curie (following a longstanding tradition for naming elements) named her first discovery after Poland (polonium), while her student Marguerite Perey later honored France with francium. Nations competed for scarce radioactive resources and promoted indigenous industries. Institutions for radioactivity research were created along national lines to bolster both prestige and economic benefits.

Some scientists had discerned national differences in scientific styles, tracing them to differences in thinking styles. These analyses served as much for propaganda as for enlightenment, since proponents generally found reasons to prefer the style attributed to their own nation. The generalizations about national differences, though useful as a first approximation, were riddled with exceptions.

In spite of competitive attitudes that predated the twentieth century, scientific cooperation across national lines was the rule before World War I. American industrialist Andrew Carnegie funded scholarships for Marie Curie's and Soddy's laboratories. Radioactivity's leaders welcomed students from far-flung lands without prejudice (though Marie Curie had a special fondness for Polish students). Scientists prided themselves on being "apolitical," a term which would be considered self-serving a few decades later, during the Nazi era.

After World War I erupted, nationalism openly threatened the ideal of an independent, objective science that stood above petty personal concerns and partisan politics. Ninety-three prominent German academics signed a contentious proclamation supporting their fatherland. Many of their colleagues outside Germany were dismayed and disappointed. From their perspective, the signers had exchanged the internationalist ideal for a narrow jingoism.

The allies likewise took measures against their newly designated enemies. For instance, the International Commission on Atomic Weights refused to accept German members, forcing them to create a German commission. Several British scientists of German ancestry were harassed and pressured to resign. Emotions flared on both sides of the English Channel.

Afterwards the victors rearranged Europe's map along ethnic lines by carving up Austria-Hungary, the surviving remnant of the once powerful Roman Empire that had incorporated diverse cultures. The First World War and its aftermath exacerbated old animosities and created new ones, setting in motion events which eventually led to a second world war. Meyer commented bitterly on the situation in 1920: "During the war we were at peace and friendship with Bohemia, Poland and all the other inventions; after the war they are constructed as new foes."[7]

In spite of rampant nationalism, scientific friendships often survived the strain of war. Soddy sent a lead sample to Vienna during the war so that Hönigschmid could determine its atomic weight. Robert W. Lawson had a relatively easy time at the Vienna Radium Institute, where (after two brief arrests) he was allowed to continue his research undisturbed through Meyer's intervention. Meyer attested to similar favors that Rutherford granted to his countrymen stranded in England.[8]

Rutherford's student James Chadwick had the misfortune of being at Berlin's Physical Technical Institute working with Hans Geiger when war broke out in 1914. Since he was called up from the reserves, Geiger could not protect Chadwick, who spent the next four years under arrest as an enemy citizen. Yet, Chadwick had a certain amount of freedom and apparently was allowed to leave the prison camp for scientific purposes. He made the most of a difficult situation, reporting that "we have been making[,] borrowing[,] and buying apparatus." Together with several other prisoners, Chadwick organized a colloquium, set up a laboratory, and kept busy with experiments. In 1918 he wrote to Rutherford that he had visited several famous scientists who were "extremely willing to help and offered to lend us anything they could. In fact all kinds of people lend us apparatus."[9]

The fact that communications could cross enemy lines during the war, permitting not only letters and verbal reports but even a sample of thorium lead to pass, attests to the strength of an internationalistic ethos among scientists. One frequent intermediary was the Dutch physicist Heike Kammerlingh Onnes, famed for his low temperature research. The Nobel Prize committee upheld the scientific ideal by awarding the 1919 chemistry prize to Fritz Haber, in spite of his role in Germany's poisonous gas warfare program.

After the war Meyer wrote to Rutherford that "we felt always sure your feelings towards us, as ours towards you, could not be harmed by the psychoses of the surroundings." During the harrowing period after the harsh Treaty of Versailles when starvation threatened the losers, Meyer managed to send food to Marie Curie's family in Poland.[10] In 1921 Rutherford obtained funds from the Royal Society to purchase radium Meyer had loaned him before the war, in order to help Meyer's financially strapped Radium Institute.

Though radioactivity researchers resumed most scientific contacts and correspondence after the war, hard feelings lingered. For instance, Marie Curie was cool towards Germans who had signed the 1914 proclamation. For years she would not accept German applicants for positions at her institute. French physicists did not want to invite German scientists to international meetings, and Germany and its ally Austria were not part of the International Research Council that formed in 1919. Two years later Rutherford told Hevesy that it was still too soon for him to visit England.[11] For their part many German scientists refused to attend professional meetings. In 1926 they ignored an invitation to join the Council.

In spite of the damage to science's internationalist values, many scientists viewed the war as extraneous to science and wanted research and professional relations to be normalized as soon as possible. The Dutch broke the ice by asking several German scientists to attend a meeting in 1922.[12] Collaboration between Germany and its former enemies gradually increased, especially with England and the United States. France, which had suffered defeat in an 1870 war with Germany, had a harder time adjusting to the changed circumstances.

During the 1930s, political realities overtook scientific ideals. Like most of their compatriots, German scientists were unable or unwilling to muster effective resistance to the Third Reich. Nazi party zealots proclaimed the superiority of German culture and "Aryan science" and tried to purge the arts and sciences of supposed alien elements. Nations turned into friends and foes regardless of scientific ties, voiding the illusion of a pure, nonpartisan science. Secret research directed towards national ends, narrowly conceived as political and military, displaced the ideal of information flowing freely for the benefit of all. Radioactivity's successor, nuclear physics, was both beneficiary and tragic agent for nationalism gone amok as the world headed towards another catastrophic war.

Radioactivity and Timeless Questions

The Quest for Understanding

Nothing is new under the sun.

—*Ecclesiastes* 1:9

The human mind is restless in the face of the unknown. Whether with myths and legends or models and theories, humans constantly strive to accommodate the unfamiliar to the familiar, in hopes of being able to predict, justify, or even control the world around them. Methods may involve invoking predecessors and precedents, similarities and parallels, or structured analogies.

Radioactivity's development as a science illustrates these processes. People tried to match the new phenomenon to familiar patterns and to use it to answer fundamental questions about the nature of the universe.

The kinds of questions posed and the answers ventured illuminate the sorts of problems considered important at the turn of the century. They also reveal more universal concerns. Timeless questions about how the world works and the philosophical underpinnings of reality became part of the new science. These included ideas about change, continuity, and matter and energy.

Because of the particular time and place that radioactivity entered into history, people tried to accommodate it to scientific worldviews. Current scientific theories and models became the raw material for constructing explanations of radioactivity.

MODELS AND THEORIES FOR RADIOACTIVITY

A common way to assimilate new information is to draw analogies, for instance, that x rays resemble light or that radioactivity seems similar to phosphorescence. From these analogies scientists can make hypotheses; for example, that x rays are a form of light or that radioactivity is a type of phosphorescence.

Hypotheses often have consequences that an experimenter can test. If x rays are a kind of light, it should be possible to reflect, refract, and diffract them. If radioactivity is a type of phosphorescence, it will need a source to stimulate it. If tests confirm a hypothesis, it may be developed into a more comprehensive theory.

Sometimes scientists will make a model for a phenomenon in order to better understand it. Models can be physical devices, but more commonly they will be mental constructs based on familiar physical processes or images. Models can also be abstract, such as a mathematical function that is applied to different kinds of phenomena.

The terms *model* and *theory* are often used interchangeably when a theory is built on a physical representation. Two models (or theories) for light were known as the "wave model" and the "particle model." A model helps the scientist to visualize a problem, and usually has consequences which an experimenter can test.

For instance, if light is a wave motion, it should spread out slightly when it passes through a narrow slit.

Physicists used a variety of physical and mathematical models and more general theories to explain radioactivity, borrowing from phosphorescence, mathematics, electromagnetic theory, thermodynamics, kinetic theory, chemistry, and astronomy. Because of the circumstances of its discovery, radioactivity was first compared to phosphorescence. After the basic phosphorescence theory proved implausible, physicists still used parts of that idea. The suggestion that some external event triggered disintegration recalled a triggering model which the German physicist Philipp Lenard developed for phosphorescence. References to an internal atomic temperature to explain radioactivity paralleled the idea of luminescence temperature proposed to explain phosphorescence and fluorescence by another German physicist, Eilhard Wiedemann.[1]

The mathematicians' exponential function was used by physicists to quantify the decay of phosphorescence and fluorescence and to describe heat processes, and by chemists to describe changes in a single molecule. This function meshed nicely with graphs of decay and regeneration of radioactive substances. With Ernest Rutherford and Frederick Soddy's groundbreaking work, the exponential model became central to the theory of radioactivity.

The crowning achievements of nineteenth-century physics were James Clerk Maxwell's theory of electromagnetism and the theory of heat known as thermodynamics. Scientists tried to apply these impressive theories to radioactivity. The electromagnetic theory was a ready resource for speculating on radioactivity's cause, whether from an outside agent like an unknown radiation or from an interior process like radiation drain. Since researchers depended on electricity and magnetism to design instruments for measuring and analyzing radioactivity, it was natural to look for electromagnetic explanations.

Any theory of radioactivity would need to conform to the requirements of thermodynamics. Researchers puzzled over radioactivity's hidden energy source, wondering how to reconcile radium's behavior with established thermodynamic laws. Underlining the importance of heat theory was the attention paid to measurements of the heat released by radium. The results amazed scientists and focused attention on the mystery of radium's energy stores. Pierre Curie used thermodynamics to describe the energy changes in radioactivity, believing this was the only legitimate way to formulate a theory of radioactivity. Thermodynamics and the related kinetic theory of gases later provided models for radioactivity's randomness.

Some chemists tried to explain radioactivity by drawing analogies to processes like the breaking apart of a molecule, the rearrangement of a molecule's parts, or oxidation. All these models failed because radioactivity was not a chemical process. Unlike chemical changes, radioactivity was not affected by temperature, pressure, sunlight, concentration, solution, or any other environmental factor.

Early in the twentieth century several scientists transferred planetary models of the solar system to atoms. In these atomic models, particles, often electrons, revolved around a more massive central body in the same way that planets orbit the sun. To explain radioactivity, scientists supposed that some atoms become unstable and explode, variously releasing electrons, alpha particles, and electromagnetic energy.

Over time scientists traced radioactivity's source to deeper and deeper regions of the atom, until they reached the newly identified atomic nucleus. By the start of the First World War researchers were considering an atom with two distinct regions: the core, or nucleus, and an outer region or atmosphere. This model was based on the prototype of the earth with its atmosphere. However, no model could dispel the ultimate mystery: Why do certain atoms explode, while others do not? To this question scientists could only reply with vague and imaginative speculations.

PATTERNS IN RADIOACTIVITY'S DEVELOPMENT

Radioactivity's development followed a common pattern in the sciences. First, something unexpected appears. Scientists try to understand the new information in terms of familiar ideas and images. If these efforts fail, they eventually revise their assumptions, creating a new outlook.

Uranium's unusual behavior mystified researchers. They tried to match their results to familiar phenomena, such as phosphorescence and chemical changes. When these attempts failed, scientists looked for other explanations. Further research and discoveries led to new ideas and questions, until in the 1930s radioactivity and other developments had produced a physics and a chemistry dramatically different from their 1896 versions.

A variety of factors moved the field forward. The first investigators were researchers intrigued by the mysterious new radiation. They sought to understand uranium's powers and the rays strictly as scientific puzzles. Soon other incentives entered the picture. Radioactivity became a tool for medicine as well as for other areas of scientific research. Manufacture of consumer products and military goods boosted the market for radioactive materials. Radioactivity's commercial possibilities attracted financial support and more researchers. Radium and other radioactive substances became highly sought commodities.

Enthusiasm for radioactivity's medical and commercial potential accelerated the field's growth. The new discipline would not have progressed so quickly without social and financial support. These factors do not seem to have affected the content of theories of radio-activity, which grew out of the era's scientific ideas and practices.

Those ideas and practices produced an interesting pattern as radioactivity developed. The new science tallied an unusual number of simultaneous discoveries and near misses, often resulting in priority disputes and hard feelings. Silvanus P. Thompson and Henri Becquerel both discovered uranium's invisible rays. Friedrich Giesel, Stefan Meyer and Egon von Schweidler, and Becquerel showed that beta rays changed course in a magnetic field. Giesel discovered radium and actinium, but Marie Curie and André Debierne got to the finish line first. Both Gerhard Schmidt and Marie Curie detected thorium's radioactivity. Willy Marckwald found the substance that Marie Curie named polonium, while Julius Elster and Hans Geitel, Giesel, and Karl Andreas Hofmann and Eduard Strauss unearthed radiolead. Elster and Geitel, Otto Hahn, and Gian A. Blanc discovered radiothorium.

The teams Rutherford-Soddy and Curie-Debierne reported similar findings when they investigated emanations and decay products. Bertram B. Boltwood, Hahn, and Marckwald all discovered ionium. Strömholm, Svedberg, and Soddy recognized isotopy, and Soddy, Fajans, Alexander Russell, and Georg von Hevesy put together the relations between the type of ray emitted and the chemical nature of the resulting product.

In 1914 Otto Hönigschmid and Stephanie Horowitz, Theodore W. Richards and Max Lembert, and Maurice Curie showed that uranium produced an isotope of lead, confirming important predictions of the transmutation theory. During the war, Lise Meitner and Hahn tracked down actinium's parent, protactinium, shortly before Soddy and John Cranston reported similar results. More examples could be listed.

Simultaneous discoveries are not unusual in science. When people with similar training and assumptions using similar methods work in the same area, they sometimes make comparable discoveries and may construct similar explanations of their findings. When a field is completely new, like radioactivity, discoveries will come more quickly and more often than in an established domain. With the surge of researchers into this exciting new area, simultaneous discoveries became quite frequent.

This confluence of discoveries was sparked first by x rays, then by curiosity about radioactivity. Radioactivity's promise for medicine attracted donations for materials, equipment, and scholarships, which supported increasing numbers of scientists. These researchers shared assumptions about how the physical world operated and the methods and tools that were appropriate for investigating it. It was inevitable that some findings would be duplicated.

Many researchers were attracted by the chance to work in a wide open field. The field's newness maximized a scientist's odds to make an original and exciting contribution. After the initial rush, possibilities gradually became more limited and more difficult to achieve.

RADIOACTIVITY AND IDEAS ABOUT CHANGE

Thinkers have wondered for centuries about the role of change in the universe. Are the changes we observe around us—like growth and decline, movement, development, and chemical changes—fundamental to the universe? Or is our world intrinsically static, so that the changes we perceive are temporary, or even outright illusions?

Physics at the turn of the century could boast of two foundational principles known as the conservation laws. These laws state that the total mass in the universe and the total energy in the universe do not change. The conservation laws assume that the universe does not change fundamentally, but only undergoes temporary alterations and conversions.

Thermodynamics confounded this tidy arrangement by revealing a universal process which could not be undone. Energy is constantly being converted into heat. No matter how carefully a device or process is designed, it will always produce some heat that cannot be changed back into more useful forms of energy, like the energy to drive a machine or to power a generator. Since this change is permanent, some energy will always be wasted.

Scientists and engineers were vividly aware of the limitation that thermodynamics imposed on their creations. Industrialists valiantly tried to increase the efficiency of the era's technological marvels: the great electrical generators, lighting systems, and factory machinery. These devices were not efficient, and gains achieved were very modest. According to thermodynamics, there was no hope of completely overcoming this problem. All forms of energy will eventually devolve into heat.

Other important discoveries in the nineteenth century supported the idea that change was fundamental to the world. Geologists found evidence of change in rocks, fossils, and the layers in the earth's surface. Anthropologists and archaeologists uncovered ancient civilizations, technologies, and peoples different from themselves. Linguists and philologists analyzed languages to find how they were related to one another and had changed over time.

In the study of living things, the realization that organisms change over time became a major organizing principle. Best known as the theory of evolution proposed by the English naturalist Charles Darwin, this principle resonated with findings in many other fields. Scientists proposed theories of the evolution of the chemical elements, of the solar system, and of the universe's origin. Other individuals applied Darwin's idea of the survival of the fittest to humans and to human societies. In this atmosphere it was natural to try to apply the evolutionary principle to radioactivity.

Gustave Le Bon seized on radioactivity to bolster his idea that all matter was evolving into energy. He thought radioactivity signaled the evolution and ultimate disappearance of matter. Le Bon extended the thermodynamic concept of irreversible change from energy to matter, without any experimental proof.

More respectable scientifically were speculations that radioactivity indicated that the chemical elements were evolving. Rutherford and Soddy's theory of radioactive change claimed that certain elements transform themselves into different elements in the process of radioactivity. With a stretch of the imagination, one could imagine that all elements had evolved over time, with radioactivity propelling the process.

Though the belief that elements evolved by shedding subatomic particles fell out of favor, the evolutionary idea later reappeared as a reverse process. Mid-twentieth-century scientists proposed that chemical elements were built up from elementary particles in a primeval inferno, the Big Bang. Evolution would continue as hydrogen fused into heavier elements in stellar furnaces like our sun.

RADIOACTIVITY AND IDEAS
ABOUT MATTER AND ENERGY

What is the primary substance in nature? Philosophers through the ages have envisioned two kinds of substances, one material and the other more abstract. The material substance might fill space completely or be restricted to small volumes in space. It might be amorphous or made of particles, uniform or composed of different elements. The more abstract component of the universe was commonly some kind of energy or force, descendant of the souls and spirits which animated the universe for eons. Philosophers could combine material and abstract elements in their speculations, or propose a universe where one type was more fundamental than the other.

Matter was considered primary during the seventeenth century, when mechanics, the study of matter in motion, was the foundation of physics. That dominance changed late in the nineteenth century, when theory suggested that energy or force might be more fundamental than matter. Thermodynamics focused on energy, and electromagnetic theory stressed force fields in space.

German chemists Wilhelm Ostwald and Georg Helm proposed basing all of physics on thermodynamics and eliminating mechanical models like atoms and molecules from theory. This program, named "energetics," was not widely accepted, though many scientists believed that thermodynamics was more fundamental than mechanics.

Around the turn of the century, the option that electromagnetism was more fundamental than matter looked promising. Maxwell's equations predicted that energy and mass should be mathematically related to one another, so that objects would become heavier as they traveled at high speed. The mass increase came from the electromagnetic field and was called "electromagnetic mass." The energy from the electromagnetic field would be proportional to mc^2, the electromagnetic mass of a moving body times the speed of light squared, with coefficients of 3/4 and 1 proposed by different physicists.

Sometimes scientists preferred the formula for energy of motion (kinetic energy), $E = \frac{1}{2}mv^2$ (where m is the mass of a moving body and v its velocity), to calculate the mass change expected for a moving particle.

Some physicists suspected that matter was composed entirely of charged particles like the electron. If this were true, everything we can see and touch might be no more than a ghostlike creation of electromagnetic fields. Mass would be a special form of electromagnetic energy and would increase as velocity increased.

Experiments in Berlin with radium's high speed electrons (beta rays) by Walter Kaufmann in 1902 seemed to confirm this, at least for electrons. The beta rays gained mass dramatically at high speeds. The electron's mass, Kaufmann concluded, was entirely electromagnetic in origin. His conclusion was widely accepted.

With its prodigious and unexplained energy output, could radioactivity be a sign that matter was changing into energy? In 1904 Soddy reasoned that "the products of the disintegration of radium must possess a total mass less than that originally possessed by the radium, and a part of the energy evolved must be considered as being derived from the change of a part of the mass into energy."[2]

The next year Albert Einstein derived a relationship between energy and mass (later expressed as $E = mc^2$, where c is the speed of light). To reach this conclusion, Einstein used a different approach than his predecessors, who had relied on electromagnetic theory. Instead, he based his derivation on what he called the principle of relativity. Einstein suggested using radium to test his equation. Since radium gives out huge amounts of energy, the mass change from its disintegration might be large enough to detect.

Without knowing Einstein's ideas, C. C. Julius Precht, a German physicist teaching in Hanover, calculated from the formula for kinetic energy that any mass change in radium would be too small to detect. Later, Einstein came to the same conclusion.

"For the moment," he stated in 1910, "there is no hope of determining that variation experimentally."[3] Even less hopeful were the odds of finding a mass change in chemical reactions, according to calculations by the Berlin chemist Hans Landolt.

The idea that radioactivity's energy might come from conversion of matter tantalized a number of scientists, including Soddy, Johannes Stark, Paul Langevin, Arnold Sommerfeld, J. J. Thomson, and Richard Swinne. More interesting and relevant for most scientists was the possibility that electromagnetic mass might cause small variations in atomic weight. This topic generated interest at a conference sponsored by industrialist Ernest Solvay in 1913. Leading chemists and physicists gathered in Belgium to consider the meeting's theme, "the structure of matter."

Conference attendees based their dialogue about mass variations on traditional electromagnetic theory rather than on Einstein's work. Scientists had no reason to prefer Einstein's ideas over the familiar physics of mechanics and electromagnetism. Either way, matter remained firmly entrenched in science. Though intriguing, notions of reducing all matter to energy forms were only speculative.

RADIOACTIVITY AND IDEAS ABOUT CONTINUITY AND DISCONTINUITY

Since antiquity philosophers had wondered whether continuity or discontinuity was more fundamental in the universe. In physics, this question translated into opposing ideas about matter and energy. Are matter and energy continuously divisible, or are they made of small, distinct parts? Can they be transferred continuously, or only in specific amounts? A rough analogy can be drawn between pouring liquid into a container and pouring rice into a container. Pouring the liquid seems like a continuous process, but pouring the rice involves transferring visibly distinct quantities, since each rice grain is a separate entity. At the end of the nineteenth century most scientists believed that matter was made of discrete parts (atoms, molecules, charged particles), but that energy was a continuous substance and could be transferred in any amount.

For radioactivity, the question was whether it came from some continuous change inside the atom or from a change which occurred in discrete steps. Conventional theory favored continuity, while some evidence supported discreteness. These concepts seemed irreconcilable.

Electromagnetic theory suggested that an atom with internal moving parts should constantly send out radiation. The continuous loss of energy would make the atom unstable, so that it would become radioactive. Yet, most atoms seemed to be stable and did not display radioactivity. It was difficult to devise a model to account for this stability when the idea of energy as a continuous substance was so firmly entrenched.

A successful model for radioactivity would also need to explain its randomness. There was no way to predict when a particular atom would decay, and no way to affect the process. Randomness suggested changes which happened in discrete stages or steps inside the atom.

In retrospect, by 1900 the belief that energy behaved like a continuous substance had created an impasse in physical theory. This assumption was so ingrained that scientists did not question it. When a breakthrough came in 1901, its author introduced it as a mere mathematical contrivance to make an equation work for a problem in radiation theory.

This theoretical problem was important for practical reasons. Most of the energy used for electrical lighting was wasted as heat. The inefficiency of lighting sources had become a major economic issue, particularly in Germany where businesses had invested heavily in electrical industries. The German government targeted considerable resources towards improving the efficiency of electrical lighting.

Scientists and engineers interested in this problem hoped to find insight from a theory of what was called "blackbody radiation." This theory related the temperature of an incandescent source (like a light bulb) to the amount and wavelength of the light it produced. Researchers tried various methods to develop a mathematical equation for this relationship. However, none of the equations they devised worked for the entire spectrum.

The German physicist Max Planck puzzled over blackbody radiation for several years. As a theoretical physicist, Planck's overriding concern was to reconcile differences in radiation theories while producing an equation that agreed with experimental results. Grappling with the discrepancies between theory and experiment, Planck decided to place a mathematical restriction on the troublesome equation. To give this restriction a physical meaning, Planck proposed that, rather than being absorbed continuously in any amount, energy was absorbed only in specific amounts. He called these amounts "quanta" (singular "quantum"), using the Latin word for "amount" (which is also the source of the English word *quantity*). This assumption challenged the idea that energy was a continuous substance.

Planck was thoroughly steeped in the nineteenth-century mindset and only reluctantly became the originator of the quantum theory. More enthusiastic than Planck initially were his German-speaking colleagues Philipp Lenard, Johannes Stark, and Albert

Einstein, who quickly became advocates of Planck's new idea. These independent-minded physicists were mavericks, as most of the scientific community was slow to appreciate the quantum innovation. Planck's idea was outside the mainstream of thought. It seemed tangential to current research on thermodynamics and did not appear to affect the most popular research areas. Rather than happening in a *quantum leap* (a term later used for changes inside atoms), the quantum revolution in physics took place gradually over several decades as various scientists developed Planck's idea into a comprehensive theory of energy.

Leaders in radioactivity did not adopt quantum theory at first, partly because it was not clear how to apply it to radioactivity. Soddy, always ahead of his time, was intrigued, as was Marie Curie, while Rutherford paid little attention at first. From his point of view the quantum theory violated common sense, since he could not visualize it. When Rutherford realized that the quantum concept agreed with experiments, he accepted the theory, as did most physicists. Not until the late 1920s did the quantum become well integrated with ideas about radioactivity. By then radioactivity had melded with the newly emerging field of nuclear physics.

ETERNAL CONUNDRUMS

Change versus permanence, matter versus energy, continuity versus discontinuity? Though a particular era may opt for one or another side of these and other philosophical issues, the next generation may well favor the contrasting stance. The universe seems to be much too complex to be contained in any simple intellectual structure that humans build to explain the world.

From one ancient perspective, such contradictions are the essence of reality. The dialectical idealism of Georg Hegel and the dialectical materialism of Karl Marx are nineteenth-century incarnations of that view. In physics, the principle that light sometimes behaves like a wave and sometimes like a particle is an example of contradictions within a theory. Known as wave-particle duality, the physics community adopted this position during the 1920s.

Another contradiction in modern physics is expressed by the so-called uncertainty principle. During the 1920s, Werner Heisenberg showed that it was impossible to specify exactly where a particle was located and how much momentum it possessed at the same time. As one measurement was made more precise, the other would become more uncertain.

Radioactivity's randomness provided an early clue to the incompleteness of physical theory at the turn of the twentieth century. This perplexing behavior was absorbed into a new physics, which embraced the contradictions that earlier researchers had tried unsuccessfully to resolve. The counterintuitive idea that nature was contradictory at its core could only enhance radioactivity's appeal to the imagination.

The Imaginative Appeal
of a Discovery

The most beautiful experience we can have is the mysterious.

—*Albert Einstein*

Radioactivity appealed to the imagination well before major contradictions in physics became obvious. Enveloped in mystery, the new phenomenon seemed to be bursting with potential. This field's attraction represented something much deeper than career aspirations or intellectual curiosity. The tale of radioactivity's discovery, the allure of its secrets, and the promise of its possibilities resonated with ancient themes signifying eternal longings of the human psyche.

Desires for power and longevity, hope for healing, yearning for transcendence, identification with beauty, romance of the mysterious, and the saga of the quest were all evoked by the new discovery. These are elements of mythology, universal and enduring in the human odyssey. The mythological worldview is an imaginative, romantic vision of the world. *Romantic* can be defined as "marked by the imaginative or emotional appeal of the heroic, adventurous, remote, mysterious, and idealized characteristics of things, places, people."[1] Radioactivity called forth all of these elements.

MYTHOLOGICAL AND ROMANTIC DIMENSIONS OF RADIOACTIVITY

Radioactivity evoked a romantic vision. It emerged suddenly out of nowhere, invisible and shrouded in mystery. It produced beautiful and eerie effects, such as luminescence, color changes in gems, and unexpected chemical reactions.

Early researchers were pioneers forging a new path with unknown destinations as they created new ideas, new instruments, and new standards. The technical work itself had a heroic cast. Radioactivity entailed risks, not considered serious enough to deter experimenters who stalwartly endured radiation burns, but sufficient to remind them that they were manipulating a very powerful agent.

The heroic tale of Marie Curie struggling against all odds and the Curies' ideal marriage of kindred souls that ended in tragedy recall stories of heroism and tragedy through the ages. The Curies were part of a myth of radium, a twentieth-century variant on traditional mythological themes. Marie Curie's search for pure radium was her pursuit of the Holy Grail, a hidden, ideal prize; her life of overcoming obstacles was the mythological hero's quest. Radium became a miraculous healing agent, the elixir of life. Journalists portrayed Curie as a heroine, almost a saint.

Curie herself embodied romantic contradictions. The contrast between her diffident, frail appearance and her seemingly superhuman feats, as well as the idea that an attractive, retiring woman and mother could succeed in a heretofore male domain, whetted the public's imagination and primed the radium myth. As a young woman she had struggled with conflicting passions: love of literature and fascination for science, romantic Polish nationalism versus positivistic objectivity and scientific internationalism, idealistic hopes for society opposed to realities of oppression and the carnage of war. After she chose to focus on scientific studies and her career, and especially after her husband's death, Marie Curie took on a grave, colorless demeanor expressed in her dress and manner.

Yet Curie's romantic soul, suppressed by professional zeal and personal sorrows, burst forth when she and Pierre purified a self-luminescent radium. It was beautiful, they exclaimed. Marie Curie could hardly wait to see what polonium would look like. Later, she reminisced on those early days: "One of our joys was to go into our workroom at night; we then perceived on all sides the feebly luminous silhouettes of the bottles or capsules containing our products. It was really a lovely sight and one always new to us. The glowing tubes looked like faint, fairy lights." On another occasion Curie remarked that "a scientist in his laboratory is ... a child placed before natural phenomena which impress him like a fairy tale."[2] In the depths of her heart Marie Curie worked in a delightful, magical world.

The discovery of atomic transmutation gave contemporary imagination a mythological theme to tap with impunity, that of the philosophers' stone and related transformation myths. It was as though the long-sought key to transformation was within reach, whether of base metals into gold, the soul from an earthly to a spiritual plane, matter to energy, or death to life.

Transformation is change. The idea that radioactive bodies were continually changing and evolving energy harmonized with an ancient view that the ultimate reality is change. As the Greek philosopher Heraclitus observed, we never step into the same river twice. Ideas which place change above permanence in the order of things have traditionally been classed as romantic, whether in science, politics, or the arts and humanities.

Radioactivity encompassed extremes of time and extension both remote and mysterious. Some radioelements decayed over eons, while others would disappear before one could blink. Their rays were invisible and contained particles smaller than atoms, yet the energies involved could be gigantic. Radioactivity's energy output was prodigious, and like the magical mill that ground salt, the cornucopia, and other symbols of plentitude, it appeared to have no end. Radioactivity's secrets also seemed limitless. Just when experimenters thought they understood something about it, new and equally mysterious behaviors would appear.

Radioactivity excited a deep chord within people, resonating with mythological themes like the hero's quest and the magic elixir, source of eternal life. Medical researchers sought healing and long life from radium, unwittingly evoking the holy wells, fountain of youth, and miraculous potions of myth and legend. The excesses of entertainers, pundits, and hucksters came from their intuitive appreciation of the subject's mythological, romantic aura. "Providence gave to the hungry, manna; and for the thirsty the rod was rent, and life was saved;" gushed an advertising brochure for the Nowata Radium Sanitarium Company, a radium water establishment in Indian Territory (later Oklahoma). The hyperbole continued: "and in no less miraculous manner the dying multitudes of earth are welcomed to Nowata—the Mecca of DeLeon [Spanish explorer who searched for the fountain of youth]."[3] These entrepreneurs offered not mere humble cures, but the elixir of life itself. Figure 14-1 depicts another advertisement for that enterprise.

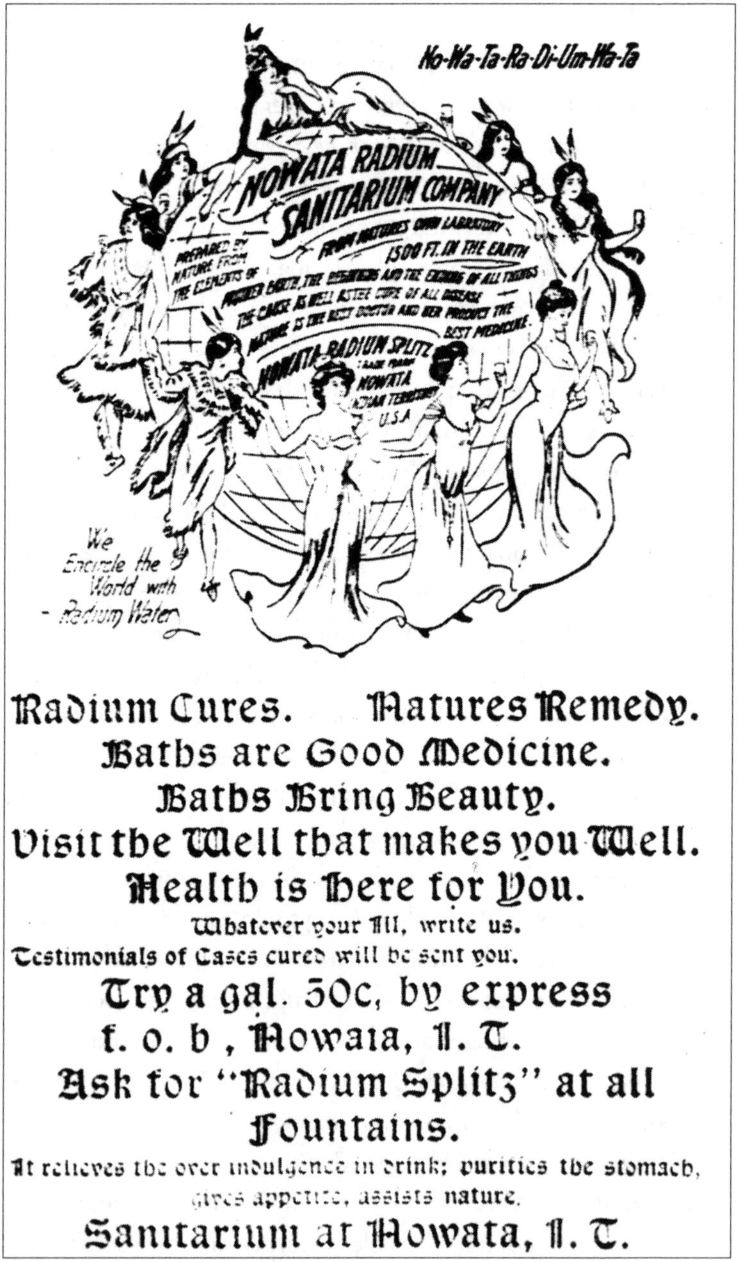

Figure 14-1. Radium water advertising brochure, c. 1905. From Mr. E. W. Hall.

In time, events dampened some of the enthusiasm. Awareness of radioactivity's hazards eroded radium's beneficent image. The first atomic bomb explosions shattered it. Radioactivity had brought to the world a destructive power of apocalyptic proportions. Physicist Robert Oppenheimer reflected on the first atomic bomb explosion by quoting from the Hindu epic Bhagavad Gita: "Now I am become Death, the destroyer of worlds."[4] A new set of images was coming to the fore.

The radium myth took on an ominous cast. No longer a magical elixir, radium became an incarnation of the poisonous apple, the fatal temptation. Science was demoted from its status as humanity's savior. The myth of Prometheus, who was severely punished for stealing fire from the gods, tempered the hopeful positivism of Marie Curie's youth. Now the legend of Doctor Faust, the man who sold his soul to the devil for knowledge, stalked science's practitioners.

AN ONGOING TASK

The noble dream of a science free of human prejudices and independent of politics and broad social concerns was ultimately unworkable. Radioactivity, at first only a scientific riddle, had effects that reached far beyond the intellectual realm. The new science's trajectory dramatically showed the futility of erecting a wall between science on the one hand, and science's applications on the other hand, with all of their accompanying political, social, and ethical implications.

This loss of innocence did not destroy belief in science's positive potential for humankind. Pierre Curie's judgement that science would provide more benefit than harm reverberated through the rest of the twentieth century and beyond, parrying with the pessimism of the doomsayers. Would the future be Hell or Eden, cataclysm or Utopia? Myth and romanticism present the extremes. Science and society continue to struggle with finding a middle ground amongst the contradictions and consequences of the powers they can wield.

APPENDIX 1

Glossary of Rays and Radiations

The terms *ray* and *radiation* were used interchangeably. After scientists learned that certain rays were particles, they often reserved the term *radiation* for electromagnetic radiation.

Alpha rays (α) Becquerel rays with a positive charge; doubly ionized helium atoms.

Becquerel rays Ionizing rays emitted from radioactive substances, including alpha, beta, and gamma rays.

Beta rays (β) Becquerel rays with a negative charge; high speed electrons.

Cathode rays Electrons ejected from the negative electrode (cathode) of a vacuum tube; sometimes used for electrons expelled from other sources.

Cosmic rays High altitude radiation; later determined to be high speed particles of extraterrestrial origin.

Delta rays (δ) Slow-moving electrons produced by alpha rays when they strike matter; a type of secondary ray.

Electromagnetic radiation Energy such as light, radio waves, and x rays, which are produced by charged particles when they change speed and/or direction.

Gamma rays (γ) Electromagnetic radiation similar to x rays but having shorter wavelengths.

Invisible light Electromagnetic radiation similar to visible light but having longer wavelengths (infrared light) or shorter wavelengths (ultraviolet light).

Primary rays Electromagnetic radiations or particles from a radioactive substance or other source that produce particles or other radiations when they strike matter.

Secondary rays Particles or electromagnetic radiation produced when primary rays strike matter.

X rays, Röntgen rays High energy electromagnetic radiation.

APPENDIX 2

Family Trees for Radioactive Elements

The *Radioactive Decay Series in 1912*

Uranium Series.

Uranium Series	Atomic weight	Weight per kilogram of Uranium	Half-value period	Rays	Range of a rays 15° C.
Ur { Uranium 1 ↓ Uranium 2	238·5	10^6 mg.	5×10^9 yrs.	a	2·5 cms.
	234·5	196 ,, (?)	10^6 yrs. (?)	a	2·9 ,,
Ur Y ↓	230·5 (?)	8×10^{-7} mg.	1·5 days	β	—
Uranium X ↓	230·5	$1·3 \times 10^{-5}$,,	24·6 ,,	$\beta + \gamma$	—
Ionium ↓	230·5	39 mg. (?)	2×10^5 yrs. (?)	a	3·00 ,,
Radium	226·5	0˙34 ,,	2000 yrs.	a	3·30 ,,

Radium Series.

Radium Series	Atomic weight	Weight per gram of radium	Half-value period	Radiation	Range of a rays at 15° C.
Radium	226	1 gr.	2000 yrs.	$a + \text{slow } \beta$	3·30 cms.
↓ Ra. Emanation	222	5.7×10^{-6} gr.	3·85 days	a	4·16 „
↓ Radium A	218	3.1×10^{-9} „	3·0 mins.	a	4·75 „
↓ Radium B	214	2.7×10^{-8} „	26·8 mins.	$\beta + \gamma$	—
↓ Radium C	214	2.0×10^{-8} „	19·5 mins.	$a + \beta + \gamma$	6·57 „
↘ Ra. C_2	—	—	1·4 mins.	β	—
Radium D	210	8.6×10^{-3} „	16·5 yrs.	slow β	—
↓ Radium E	210	7.1×10^{-6} „	5·0 days	$\beta + \gamma$	—
↓ Radium F	210	1.9×10^{-4} „	136 days	a	3·77 „

Actinium Series.

Actinium Series	Atomic weight	Half-value period	Radiation	Range of a rays at 15° C.
Actinium ...	A	?	rayless	—
↓ Radio-actinium	A	19·5 days	$a + \beta$	4·60 cms.
↓ Actinium X ...	$A - 4$	10·2 days	a	4·40 „
↓ Emanation ...	$A - 8$	3·9 secs.	a	5·70 „
↓ Actinium A ...	$A - 12$	·002 sec.	a	6·50 „
↓ Actinium B ...	$A - 16$	36 mins.	slow β	—
↓ Actinium C ...	$A - 16$	2·1 mins.	a	5·40 „
↓ Actinium D ...	$A - 20$	4·71 mins.	$\beta + \gamma$	—

Thorium Series.

Thorium Series	Atomic weight	Weight per 10^6 grams thorium	Half-value period	Radiation	Range a rays 15° C.
Thorium ... ↓	232	10^9 mg.	1.3×10^{10} yrs.	a	2.72 cms.
Mesothorium 1 ↓	228	0.42 mg.	5.5 yrs.	rayless	—
Mesothorium 2 ↓	228	5.2×10^{-5} mg.	6.2 hrs.	$\beta + \gamma$	—
Radiothorium ↓	228	0.15 „	2 yrs.	a	3.87 „
Thorium X ... ↓	224	7.4×10^{-4} „	3.65 days	$a + \beta$	4.3 „
Th. Emanation ↓	220	1.2×10^{-7} „	54 secs.	a	5.0 „
Thorium A ... ↓	216	3.1×10^{-10} „	0.14 secs.	a	5.7 „
Thorium B ... ↓	212	8.5×10^{-5} „	10.6 hrs.	$\beta + \gamma$	—
Thorium C ... β↙ ↓ a	212	7.9×10^{-6} „	60 mins.	$a + \beta$	4.8 „
Th. C_2 ↓	212	—	very short (?)	a	8.6 „
Thorium D ...	208	1.3×10^{-7} „	3.1 mins.	$\beta + \gamma$	—

Source: From Ernest Rutherford, *Radioactive Substances and their Radiations* (Cambridge: Cambridge University Press, 1913), 468, 518, 533, 552. Reproduced with permission of Ernest Rutherford's family.

The Radioactive Decay Series According to Modern Data

1. The Uranium Family

Radioelement and Rays	Half life (years, days, hours, minutes, seconds)
Uranium I	4,500,000,000 y
↓ α	
Uranium X_1 (Thorium 234)	24 d
↓ β	
Uranium X_2 (Protactinium 234)	1.2 m
↓ β	

(continued)

1. The Uranium Family—(*continued*)

Radioelement and Rays		*Half life* (*years, days, hours, minutes, seconds*)
Uranium II (Uranium 234)		240,000 y
↓α		
Ionium (Thorium 230)		77,000 y
↓α		
Radium (Radium 226)		1,600 y
↓α		
Radium emanation (Radon 222)		3.8 d
↓α		
Radium A (Polonium 218)		3.1 m
↓α or ↘β		
Radium B (Lead 214)	Astatine (Astatine 218)	27 m, 2 s
↓β	↙α	
Radium C (Bismuth 214)		20 m
↓β or ↘α		
Radium C' (Polonium 214)	Radium C" (Thallium 210)	0.00016 s, 1.3 m
↓α	↙β	
Radium D (Lead 210)		22 y
↓β		
Radium E (Bismuth 210)		5.0 d
↓β or ↘α		
Radium F (Polonium 210)	Thallium (Thallium 206)	140 d, 4.2 m
↓α	↙β	
Radium G (Lead 206)		Not radioactive

2. The Actinium Family

Radioelement and Rays	Half life (years, days, hours, minutes, seconds)
Actinouranium (Uranium 235)	710,000,000 y
$\downarrow \alpha$	
Uranium Y (Thorium 231)	26 h
$\downarrow \beta$	
Protactinium (Protactinium 231)	33,000 y
$\downarrow \alpha$	
Actinium (Actinium 227)	22 y
$\downarrow \beta$ or $\searrow \alpha$	
Radioactinium Actinium K (Thorium 227) (Francium 223)	19 d, 22m
$\downarrow \alpha$ $\swarrow \beta$	
Actinium X (Radium 223)	11 d
$\downarrow a$	
Actinium emanation (Radon 219)	4.0 s
$\downarrow \alpha$	
Actinium A (Polonium 215)	0.0018 s
$\downarrow \alpha$ or $\searrow \beta$	
Actinium B Astatine (Lead 211) (Astatine 215)	36 m, 0.0001 s
$\downarrow \beta$ $\swarrow \alpha$	
Actinium C (Bismuth 211)	2.1 m
$\downarrow \beta$ $\searrow \alpha$	
Actinium C' Actinium C" (Polonium 211) (Thallium 207)	0.005 s, 4.8 m
$\downarrow \alpha$ $\swarrow \beta$	
Actinium D (Lead 207)	Not radioactive

3. The Thorium Family

Radioelement and Rays	Half life (years, days, hours, minutes, seconds)
Thorium (Thorium 232)	14,000,000,000 y
↓ α	
Mesothorium I (Radium 228)	5.8 y
↓ β	
Mesothorium II (Actinium 228)	6.1 h
↓ β	
Radiothorium (Thorium 228)	1.9 y
↓ α	
Thorium X (Radium 224)	3.7 d
↓ α	
Thorium emanation (Radon 220)	56 s
↓ α	
Thorium A (Polonium 216)	0.15 s
↓ α or ↘ β	
Thorium B Astatine (Lead 212) (Astatine 216)	11 h, 0.0003 s
↓ β ↙ α	
Thorium C (Bismuth 212)	61 m
↓ β or ↘ α	
Thorium C' Thorium C" (Polonium 212) (Thallium 208)	0.0000003 s, 3.1 m
↓ α ↙ β	
Thorium D (Lead 208)	Not radioactive

Sources: Samuel Glasstone, *Sourcebook on Atomic Energy*. Princeton, NJ: D. Van Nostrand, 1950.

    Argonne National Laboratory, EVS, *Human Health Fact Sheet*. Argonne National Laboratory (Illinois), 2005.

APPENDIX 3

Radioactivity's Elusive Cause

From its discovery in 1896, radioactivity mystified researchers because they could not find an energy source for the exploding atoms. No matter what sources they tested and which novel methods they tried, they could not change the process of radioactivity. They failed because radioactive elements do not use energy from outside themselves to disintegrate.

An atom needs energy in order to disintegrate. The energy that allows certain atoms to spontaneously decay and release excess energy comes from inside these atoms. Atoms of naturally radioactive heavy elements can disintegrate because they are not stable. They have a tendency to come apart because they contain large numbers of mutually repelling positive charges (protons) within their nuclei.

The force that causes like charges to repel and unlike charges to attract is called the *electrostatic force*. The electrostatic force creates energy when charged particles push themselves apart or pull themselves together. It takes a good deal of energy to overcome the electrostatic forces inside heavy nuclei and keep the protons from flying apart.

That energy comes from a force inside the nucleus called the *strong force*. The strong force holds protons and neutrons together. If the strong force holding these particles together is greater than the electrostatic force pushing the protons apart, the nucleus will be stable.

If an unstable nucleus casts off an alpha particle, the new element formed may send gamma rays out from its nucleus as it assumes a more stable configuration. The gamma radiation represents the energy difference between different energy levels of the nucleus.

Researchers were puzzled to find cases where alpha decay occurred even though the electrostatic forces were slightly less than the strong forces within a nucleus. They could not understand how an alpha particle could escape the nucleus if it did not have enough energy to overcome the forces that kept the protons and neutrons together. These examples seemed to violate the conservation of energy.

A theory developed in the 1920s called *wave mechanics* showed that, although high energy alpha particles are most likely to be able to escape a nucleus, even low energy particles can do so. Whether or not a particular particle escapes is a matter of probability, and the probability for escaping never goes to zero. These probabilities, calculated by using wave mechanics, determine how long it takes, on average, for atoms of a particular radioelement to decay. Some radioelements have average lives of less than a second, while others have lifetimes of millions or even billions of years.

In the 1930s physicists learned that neutrons themselves can disintegrate. This process produces beta particles. The force involved in beta particle emission was later named the *weak force*, in contrast to the *strong force* that holds nuclei together. In beta decay a neutron transmutes into a proton, an electron (beta particle), and an uncharged, nearly massless particle called an anti-neutrino. Beta particle decay is also governed by laws of probability. After a beta particle is ejected, a new nucleus is formed, which may emit gamma rays.

During the first years of radioactivity scientists believed that deterministic, mechanical causes lay underneath the probabilistic equations that described radioactivity. They reasoned that although probability theory could describe radioactivity, these abstract equations did not explain it. Radioactivity's leaders assumed that a mechanical cause or causes for radioactivity might be found in the future.

After quantum mechanics was developed and interpreted in terms of probabilities, many scientists decided that the probabilistic equations were themselves the explanation of radioactivity. There was no deeper cause for the exploding atoms. At the subatomic level, Nature was governed by Chance.

APPENDIX 4

Nobel Prize Winners Included in This Book

Names are listed according to *Nobel Lectures. Physics, 1901–1921; 1922–1941; 1942–1962;* and *Nobel Lectures. Chemistry, 1901–1921; 1922–1941; 1942–1962* (Amsterdam: Elsevier, 1964–).

Physics

1901	Wilhelm Conrad Röntgen
1902	Pieter Zeeman
1903	Antoine Henri Becquerel, Pierre Curie, and Marie Skłodowska-Curie
1904	Lord Rayleigh (John William Strutt)
1905	Philipp Eduard Anton von Lenard
1906	Joseph John Thomson
1911	Wilhelm Wien
1913	Heike Kamerlingh Onnes
1914	Max von Laue
1915	William Henry Bragg and William Lawrence Bragg
1917	Charles Glover Barkla
1918	Max Planck
1919	Johannes Stark
1921	Albert Einstein
1922	Niels Bohr
1926	Jean Baptiste Perrin
1927	Charles Thomson Rees Wilson
1935	James Chadwick
1936	Victor Franz Hess

Chemistry

1904	Sir William Ramsay
1908	Ernest Rutherford
1909	Wilhelm Ostwald
1911	Marie Skłodowska Curie
1914	Theodore William Richards
1921	Frederick Soddy
1935	Jean Frédéric Joliot and Irène Joliot-Curie
1943	George de Hevesy (Georg von Hevesy)
1944	Otto Hahn

APPENDIX 5

Radioactivity's Web of Influence (Illustration by author)

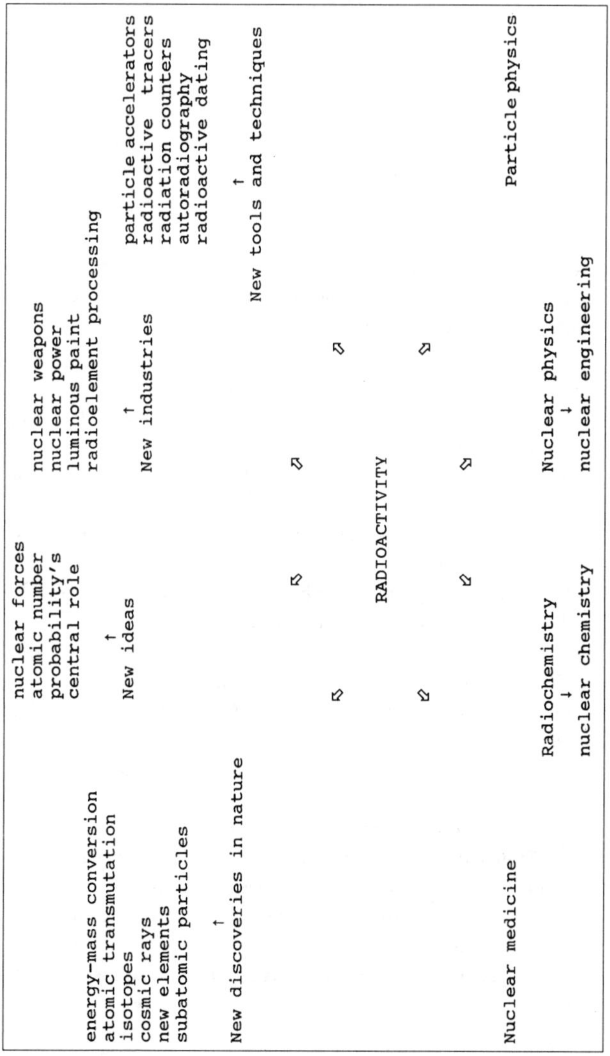

APPENDIX 6

Timeline

Discoveries are complex events that usually involve multiple persons, events, and interpretations. I have used the term here in its simplest sense, reflecting the most common interpretation of the historical record.

1789	Uranium is discovered by Martin Klaproth
1855	Johann Geissler invents a mercury vacuum pump and develops specialized glass tubes for viewing electrical discharges
	Heinrich Rühmkorff invents an induction coil, or transformer, that can produce high voltages
1859	Gustav Kirchhoff and Robert Bunsen show that spectral lines can be used to identify chemical elements
1860	James Clerk Maxwell and Ludwig Boltzmann independently develop a theory for behavior of gas molecules
1864	Maxwell publishes his equations for electromagnetism
1865	Hermann Sprengel improves the mercury vacuum pump
1869	Johann Hittorf identifies cathode rays, named by Eugen Goldstein in 1876
	Dmitri Mendeleyev publishes his first version of the periodic table of the elements
1873	Maxwell predicts the existence of radio waves
1881	J.J. Thomson proposes the idea of electromagnetic mass
1895	Wilhelm Röntgen discovers invisible penetrating rays, or "X" rays
1896	Henri Becquerel discovers invisible penetrating rays from uranium; Silvanus P. Thompson makes a similar discovery
	Pieter Zeeman observes the splitting of spectral lines by a magnetic field

1897   Emil Wiechert and J. J. Thomson independently identify cathode rays as negatively charged particles much smaller than an atom (later called electrons); Walter Kaufmann obtains similar results but does not draw this conclusion

1898   Marie and Pierre Curie coin the terms "radio-active" and "radio-activity"

Marie and Pierre Curie discover polonium; together with Gustave Bémont, they discover radium

Gerhard Carl Schmidt discovers that thorium is radioactive

Friedrich Giesel begins producing radium

Ernest Rutherford identifies two components in the uranium rays, which he names "alpha" and "beta"

1899   Rutherford discovers the thorium emanation; Pierre and Marie Curies observe "induced radioactivity"

J. J. Thomson measures the charge on the electron, proving that it is a subatomic particle

Giesel reports that polonium's activity is not permanent

Giesel and the team of Stefan Meyer and Egon von Schweidler, closely followed by Becquerel, deflect beta particles in a magnetic field, proving that they are charged particles rather than a type of x ray

André Debierne discovers actinium

1900   Paul Villard discovers gamma rays

Ernst Dorn deflects beta rays in an electric field

Dorn discovers radium emanation

Rutherford applies the exponential function to radioactivity

Max Planck introduces the idea of the energy quantum

Friedrich Otto Walkoff reports that radium burns skin

1901   Röntgen receives the first Nobel Prize for Physics

1902   Rutherford and Frederick Soddy publish their transformation theory of radioactivity

Rutherford identifies the alpha particle

Giesel begins marketing radium

Marie Curie determines the atomic weight of radium

Walter Kaufmann shows that beta ray electrons increase in mass as their velocity increases

1903   Pierre Curie and Albert Laborde measure the heat output of radium

Soddy and William Ramsay show that radium produces helium

Franz Himstedt identifies radium emanation (radon) in well water and petroleum

Sir William Crookes and research partners Julius Elster and Hans Geitel independently observe scintillations when alpha particles strike zinc sulfide screens

Becquerel and the Curies receive the Nobel Prize for Physics

1904   Two journals dedicated to radioactivity begin publication, *Le Radium* and *Jahrbuch der Radioaktivität und Elektronik*

Armet de Lisle begins commercial production of radium near Paris

1905   Robert J. Strutt uses helium for radioactive dating

Schweidler applies probability theory to radioactivity, allowing scientists to confirm that radioactivity is a random process

First death attributed to radioactivity

1906   Giesel begins marketing luminous paint containing radium

Charles Barkla discovers characteristic x rays

Norman J. Campbell and Albert B. Wood find that potassium and rubidium are radioactive; this property was later traced to radioactive isotopes of these elements

Pierre Curie (1859–1906)

1908   Hans Geiger invents a radiation counter

Rutherford receives the Nobel Prize for Chemistry

Henri Becquerel (1852–1908)

1910   The International Radium Standards Committee is formed

The Vienna Institute for Radium Research opens

Yale institutes the first academic position for radiochemistry; Bertram B. Boltwood is its first occupant

1911   Geiger and John M. Nuttall develop a law relating rate of decay to alpha particle range

Rutherford publishes his scattering theory, proposing a planetary atom with a central core

Soddy formulates the concept of isotopes

Antonius van den Broek concludes that the periodic table should be ordered by atomic number, rather than by atomic weight

C. T. R. Wilson develops a cloud chamber for viewing the path of a charged particle

Marie Curie receives the Nobel Prize for Chemistry

1912   The Kaiser Wilhelm Institute for Chemistry opens with a radioactivity section

An international standard for radium is adopted

Victor Hess confirms the existence of a high altitude radiation, later called cosmic rays

The radioactive displacement laws are formulated; Kasimir Fajans, Soddy, Georg von Hevesy, Alexander S. Russell, and Alexander Fleck are instrumental

Walter Friedrich and Paul Knipping show that x rays behave like waves

Marie Curie foreshadows the wave mechanical interpretation of radioactivity

1913	Bohr publishes an atomic theory linking quantum theory and Rutherford's nuclear atom
	Francis W. Aston identifies isotopes in neon, a nonradioactive element
	Georg von Hevesy and Friedrich Paneth show that radioactive lead can be used as a chemical indicator, or tracer
	The Polish Radium Institute opens
	William H. Bragg and William L. Bragg measure the wavelengths of x rays
	Rutherford and Edward N. da Costa Andrade demonstrate interference of gamma rays, confirming their wave nature
1914	Otto Hönigschmid and Stephanie Horowitz show that lead from uranium weighs less than ordinary lead
	Henry G. J. Moseley finds a mathematical relation between atomic number and the frequency of characteristic x rays
	The Paris Institute of Radium opens
	Debierne introduces the idea of nuclear surface tension
1914–1918	World War I
1915	W. H. Bragg and W. L. Bragg receive the Nobel Prize for Physics
	Henry G. J. Moseley (1887–1915)
1916	Silvanus P. Thompson (1851–1916)
	Sir William Ramsay (1852–1916)
	Ernst Dorn (1848–1916)
1917	Beginnings of dial painting with paint containing radium
1919	Aston develops a mass spectrograph to separate isotopes
	Rutherford disintegrates nitrogen atoms by bombarding them with alpha particles
	Sir William Crookes (1832–1919)
1920	Julius Elster (1854–1920)
1921	Soddy receives the Nobel Prize for Chemistry
1922	The Belgian Mining Society begins producing radium from African pitchblende
	First death of a dial painter
	Aston receives the Nobel Prize for Chemistry
1923	Georg von Hevesy uses radioactive lead as a tracer to study plant metabolism
	Louis-Victor de Broglie proposes that particles can behave like waves
	Hans Geitel (1855–1923)
	Wilhelm Conrad Röntgen (1845–1923)
1925	Werner Heisenberg publishes his first paper on quantum mechanics

1926	Erwin Schrödinger applies de Broglie's theory to the atom, creating a theory known as wave mechanics
	Antonius van den Broek (1870–1926)
1927	Friedrich Giesel (1852–1927)
	Bertram B. Boltwood (1870–1927)
	C.T.R. Wilson receives the Nobel Prize for Physics
1928	Geiger and Walther Müller develop a sensitive radiation counter
	George Gamow and the team of Ronald W. Gurney and Edward U. Condon independently explain alpha particle decay using wave mechanics
	Georges Sagnac (1869–1928)
1932	James Chadwick discovers the neutron
	Carl Anderson discovers the positive electron (positron) in cosmic rays
	Radium is produced from Canadian pitchblende
	The American Medical Association withdraws approval for internal use of radium
1933	Enrico Fermi presents a theory of beta particle decay
1934	Irène Curie and Frédéric Joliot discover artificial radioactivity
	Marie Skłodowska Curie (1867–1934)
	Paul Villard (1860–1934)
1935	James Chadwick receives the Nobel Prize for Physics
1936	Victor Hess receives the Nobel Prize for Physics
1937	Ernest Rutherford (1871–1937)
1938	Otto Hahn and Fritz Strassmann announce the production of barium from uranium; Lise Meitner and Otto Frisch explain the process, named "fission"
1939–1945	World War II
1940	Joseph John (J. J.) Thomson (1856–1940)
1942	William H. Bragg (1862–1942)
	Willy Marckwald (1864–1952)
1943	Hevesy receives the Nobel Prize for Chemistry
1944	Otto Hahn receives the Nobel Prize for Chemistry
1945	The United States drops two fission bombs over Japan
	Beginnings of the Cold War
	Hans Geiger (1882–1945)
	Francis W. Aston (1877–1945)
	Otto Hönigschmid (1878–1945)
1948	Egon Ritter von Schweidler (1873–1948)
1949	Stefan Meyer (1872–1949)
	André-Louis Debierne (1874–1949)
	Gerhard Carl Schmidt (1865–1949)
1956	Frederick Soddy (1877–1956)

1962   Niels Bohr (1885–1962)
1966   Georg von Hevesy (1885–1966)
1968   Otto Hahn (1879–1968)
        Lise Meitner (1878–1968)
        Alexander Fleck (1889–1968)
1972   Alexander S. Russell (1888–1972)
1975   Kasimir Fajans (1887–1975)

NOTES

In order to avoid an excessive number of notes, I have limited citations to direct quotes, clarifications, and a few additional sources. References are abbreviated only when they are cited more than once in the same chapter.

Chapter 1

1. Some cathode rays can pass through thin aluminum foil.

Chapter 2

1. Marie Curie, *Pierre Curie*, trans. Charlotte and Vernon Kellog (New York: Dover, 1963), 50; Eve Curie, *Madame Curie*, trans. Vincent Sheean (Garden City, NY: Doubleday, Doran & Company, 1937), 287.
2. M. Curie, *Pierre Curie*, 34.
3. Friedrich Giesel, "Ueber Radium und radioaktive Stoffe," *Deutsche Chemische Gesellschaft: Berichte* 35:3 (1902): 3608–11, on 3609; "Über Radium und Polonium," *Physikalische Zeitschrift* 1 (1899): 16–17; "Einiges über das Verhalten des radioaktiven Baryts und über Polonium," *Annalen der Physik und Chemie* 69 (1899): 91–94. Giesel described the light as bluish or blue-green. Industrialist Eugen de Haën reported the self-luminosity shortly before Giesel; see his "Ueber eine radioaktive Substanz," *Annalen der Physik und Chemie* 68 (1899): 902. See also P. Adloff and H. J. MacCordich, "The Dawn of Radiochemistry," *Radiochimica Acta* 70/71 (1995): 13–22.

4. Friedrich Giesel to the Curies, 6 January 1900, Archives Institut du Radium—fonds Marie Curie, Musée Curie, Institut Curie; Giesel, "Ueber Radium und radioaktive Stoffe."

5. Friedrich Giesel to the Curies, 22 December 1899, Archives Institut du Radium—fonds Marie Curie, Musée Curie, Institut Curie, quoted in Marjorie Malley, "The Discovery of the Beta Particle," *American Journal of Physics* 39 (1971): 1459, n. 5. Author's translation.

6. From Ostwald's autobiography; quoted by Robert Reid, *Marie Curie* (New York: New American Library, 1974), 76.

7. Polonium proved more refractory. Because of its brief half life, Marie Curie was not able to isolate a weighable quantity.

8. *Nobel Lectures. Physics, 1901–1921* (Amsterdam: Elsevier, 1967), vol. 1, 45.

Chapter 3

1. Becquerel was reluctant to totally discard his hypothesis, arguing that the rays might contain a small amount of light.

2. Elster and Geitel's tests with pressure changes and heat showed slight changes, which they did not consider significant.

3. The depth was 852 meters, nearly 3,000 feet.

4. Pierre and Marie Curie, "Les nouvelles substances radioactives et les rayons qu'elles émmittent," *Rapports présentes au congrès international de physique*, 3 (Paris: Gauthier-Villars, 1900), 79–114, on 114.

5. Sir William Crookes, "Radio-activity of Uranium," *Proceedings of the Royal Society of London A* 66 (10 May 1900): 409–422.

6. Muriel Howarth, *Pioneer Research on the Atom* (London: New World Publications, 1958), 83–4.

7. The actual decay sequence is more complex.

8. Ernest Rutherford and Frederick Soddy, "The Radioactivity of Thorium Compounds. II," *Transactions of the Chemical Society of London* 81 (15 May 1902): 837–60, in *The Collected Papers of Lord Rutherford of Nelson*, vol. 1 (London: George Allen and Unwin, 1962), 435–56, on 455.

9. Frederick Soddy, "Alchemy and Chemistry," Soddy Papers, Bodleian Library, Oxford University, file 1; Frederick Soddy to Ernest Rutherford, 7 August 1903, Rutherford Papers, letter S99. Letter courtesy of the Syndics of Cambridge University Library.

10. Ernest Rutherford and Frederick Soddy, "Note on the Condensation Points of the Thorium and Radium Emanations," Proceedings of the Chemical Society of London (1902): 219–20, in *The Collected Papers of Lord Rutherford of Nelson*, vol. 1, 528.

11. Frederick Soddy to Ernest Rutherford, 12 July 1902, Soddy Papers, Bodleian Library, Oxford University.

12. Friedrich Giesel to the Curies, 12 January 1900, Archives Institut du Radium—fonds Marie Curie, Musée Curie, Institut Curie; Friedrich Giesel to Carl Runge, 25 November 1899, Deutsches Museum Archiv HS 1948-52, Munich.

13. Debierne later earned a doctorate in physics.

14. Pierre and Marie Curie, "Sur les corps radio-actifs," *Comptes Rendus* 134 (13 January 1902): 85-87.

15. Pierre Curie, "Sur la radioactivité induite et sur l'émanation du radium," *Comptes Rendus* 136 (26 January 1903): 223-26, on 226; Pierre Curie, "Sur la constante de temps caractéristique de la disparition de la radioactivité induite par le radium dans une enceinte fermée," *Comptes Rendus* 135 (17 November 1902): 857-59, on 859.

16. Paul Langevin, "Pierre Curie," *Revue du Mois* 2 (July-December 1906): 5-36, on 27.

17. Pierre Duhem, *The Aim and Structure of Physical Theory*, trans. Philip P. Wiener (New York: Atheneum, 1962), 70-71.

18. Henry G. J. Moseley to E. Rutherford, 5 June [1914], and Ernest Rutherford to W. H. Bragg, 20 December 1911, cited in Arthur S. Eve, *Rutherford* (New York: Macmillan, 1939), 237 and 208.

19. Pierre Curie, "Recherches récentes sur la radioactivité," *Journal de chimie physique* 1 (1903): 409-49, on 446-47.

20. Notes on radioactivity for his course at the Sorbonne, 1904-05 and 1905-06, Archives Pierre et Marie Curie, Bibliothèque Nationale de France.

21. Frederick Soddy to Ernest Rutherford, 19 February 1903, Rutherford Papers, letter S93; Clemens Winkler, "Radio-activity and Matter," *Chemical News* 89 (17 June 1904): 289-91, on 290; Joseph Larmor to Ernest Rutherford, 3 October 1903, Rutherford Papers, letter L22. Letters courtesy of the Syndics of Cambridge University Library.

22. Meaning a dreamlike succession of complex shifting images or an optical projector arranged to produce this effect. Marie Curie, "Les radio-éléments et leur classification," *Revue du Mois* 18 (10 July 1914), 5-41, in *Oeuvres de Marie Skłodowska Curie* (Warsaw: Polish Academy of Sciences, 1954), 472-93, on 472-73.

23. Arthur Smithells, Presidential Address to Section B, *Reports of the British Association for the Advancement of Science* 77 (August 1907): 469-79, on 477; Muriel Howarth, *Atomic Transmutation* (London: New World Publications, 1953), 86-87.

24. Willy Marckwald, "Die Radioaktivität," *Deutsche Chemische Gesellschaft: Berichte* 41:2 (1908): 1524-61, on 1536.

25. Eve, *Rutherford*, 374.

26. Pierre Curie, "Radioactive Substances, Especially Radium," *Nobel Lectures. Physics, 1901-1921*, 73-78, on 78.

27. Frederick Soddy, *The Interpretation of Radium*, 1908, cited in Howarth, *Pioneer Research on the Atom*, 122.

28. George Jaffe, "Recollections of Three Great Laboratories," *Journal of Chemical Education* 29 (1952): 230–38, on 238.
29. *Nobel Lectures. Chemistry 1901–1921* (Amsterdam: Elsevier, 1966), 197.
30. These experiments took place before Rutherford deflected alpha rays, in 1902.
31. Eve, *Rutherford*, 183; *Nobel Lectures. Chemistry*, 123.

Chapter 4

1. Cosmic rays were named by Robert A. Millikan in 1925. See Samuel Glasstone, *Sourcebook on Atomic Energy* (Princeton, NJ: D. Van Nostrand, 1950), 476.

Chapter 5

1. Ernest Rutherford, *Radioactive Substances and Their Radiations* (Cambridge: Cambridge University Press, 1913), 622.
2. Frederick Soddy, "Multiple Atomic Disintegration. A Suggestion in Radioactive Theory," *Philosophical Magazine* 18 (November 1909): 739–44, on 739.
3. Frederick Soddy, *Radio-activity* (New York: D. Van Nostrand, 1904), 179.
4. Marie Curie, "Sur la loi fondamentale des transformations radioactives," in *La structure de la matière*, Institut de Physique Solvay, 1913 (Paris: Gauthier-Villars, 1921), 66–71, on 70; André Debierne, "Considérations sur le méchanisme des transformations radioactives et la constitution des atomes," *Annales de physique* 4 (1916): 323–45, on 345.
5. M. Curie, "Sur la loi fondamentale," 71.

Chapter 6

1. Arthur Smithells, Presidential Address to Section B, *Reports of the British Association for the Advancement of Science* 77 (August 1907): 469–79, on 477.
2. Ernest Rutherford, "The Succession of Changes in Radioactive Bodies," *Philosophical Transactions of the Royal Society of London* 204A (1904): 169–219, in *Collected Papers of Lord Rutherford of Nelson* (London: George Allen and Unwin, 1962), on 674.
3. Marie Curie, "Radium and New Concepts in Chemistry," *Nobel Lectures. Chemistry 1901–1921*, 1911 lecture (Amsterdam: Elsevier, 1966), 202–12, on 211.
4. Georg von Hevesy, "A Scientific Career," *Perspectives in Biology and Medicine* 1 (1958): 345–65, on 349n.

5. Bertram B. Boltwood to Ernest Rutherford, 22 September 1905, cited in Lawrence Badash, *Radioactivity in America* (Baltimore: Johns Hopkins University Press, 1979), 113.

6. Robert Dekosky, "Spectroscopy and the Elements in the Late Nineteenth Century: The Work of Sir William Crookes." *British Journal for the History of Science* 6 (1972–73): 400–23, on 422. The empty places in Soddy's table were eventually filled by technicium, rhenium, astatine, and francium.

7. Many radioelements could decay by losing either an alpha particle or a beta particle. Though suspected earlier, the fact that a radioelement could decay in more than one way (called *lateral disintegration*, or *branching*) was confirmed during the period 1909–1911. After the branching sequences (and a few more unusual cases) were identified, the analogies between the decay series became complete.

8. Willy Marckwald, "Zur Kenntnis des Mesothoriums," *Deutsche Chemische Gesellschaft: Berichte* 43:3 (1910): 3420–22, on 3421.

9. Oswald Göring, cited in Elizabeth Rona, *How It Came About* (Oak Ridge, TN: Oak Ridge Associated Universities, 1978), 7.

10. Frederick Soddy, *The Interpretation of Radium* (New York: G. P. Putnam's Sons, 1920), cited in Mary E. Weeks, *Discovery of the Elements* (Easton, PA: Journal of Chemical Education, 1968), 800.

11. Georg von Hevesy to Arthur S. Eve, 28 October 1937, Rutherford Papers, and Ernest Rutherford to Kasimir Fajans, 2 April 1913, Rutherford Papers, letter XF2. Courtesy of the Syndics of Cambridge University Library.

12. Frederick Soddy to Stefan Meyer, 11 July 1914, Meyer Papers, Institut für Radioaktivität und Kernphysik, Vienna.

13. Robert Merton, "The Matthew Effect in Science," *Science* 159 (January 5, 1968): 56–63. Soddy suffered from this effect. Radiochemist Friedrich Paneth commented that "To Rutherford is attributed the sole merit—illustrating the old truth, well known to students of the history of science, that great reputations tend to absorb the smaller ones." Quoted in Muriel Howarth, *Pioneer Research on the Atom* (London, New World Publications, 1953), 277. Paneth also had this problem, since he was overshadowed by Hevesy.

14. Bertram B. Boltwood, "The Origin of Radium," *Philosophical Magazine* 9 (April 1905): 599–613, on 613.

Chapter 7

1. David Wilson, *Rutherford. Simple Genius* (Cambridge, MA: MIT Press, 1983), 291.

2. Van den Broek used the German word *Ordnungszahl*, meaning "ordinal number."

3. William H. Bragg, "A Comparison of Some Forms of Electric Radiation," *Proceedings of the Royal Socviety of South Australia* 31 (7 May 1907): 79–93.
4. W. H. Bragg to Ernest Rutherford, 18 January 1913, Rutherford Papers, letter B394. Courtesy of the Syndics of Cambridge University Library.

Chapter 8

1. Ernest Rutherford to Stefan Meyer, 1920, cited in Arthur S. Eve, *Rutherford* (New York: Macmillan, 1939), 276.
2. Initially, Baumbach was not interred, due to Rutherford's intervention. His intemperate patriotic outbursts soon caused him to lose his freedom. See John B. Birks, ed., *Rutherford at Manchester* (New York: W. A. Benjamin, 1963), 137.
3. Though often called "splitting the atom," technically Rutherford had achieved artificial transmutation, not fission. In the transmutation of nitrogen a nitrogen atom absorbs an alpha particle, causing the nucleus to rearrange itself and eject a hydrogen ion (later named a *proton*). In fission, a nucleus breaks into smaller pieces. J. J. Thomson had tried earlier to transmute elements with x rays.
4. Many scientists had speculated that all elements were made of hydrogen, the lightest element.

Chapter 9

1. Summary of Pierre Curie's lecture at the Royal Institution, "Radium," *The Electrician* 51 (1903): 403–4, on 404.
2. The scintillations are caused by *triboluminescence*, a type of luminescence created by mechanical action on the crystals.
3. Stefan Meyer, "Das Spinthariskop und Ernst Mach," *Zeitschrift für Naturforschung* 5a (July 1950): 407–8.
4. Arthur S. Eve, *Rutherford* (New York: Macmillan, 1939), 328.
5. Elizabeth Rona, *How It Came About* (Oak Ridge, TN: Oak Ridge Associated Universities, 1978), 38; David Wilson, *Rutherford. Simple Genius* (Cambridge, MA: MIT Press, 1983), 573, as reported by Lord Ritchie Calder.

Chapter 10

1. Friedrich Giesel to Carl Runge, 20 November 1902, Deutsches Museum Archiv HS 1948–52, Munich. Walkoff became well-known for his research on radioactivity's effects on live tissues.
2. Marie Curie, *Pierre Curie*, trans. Charlotte and Vernon Kellog (New York: Dover, 1963), 56.

3. O. Peter Snyder and D. M. Poland, "Food Irradiation Today," 1995, http://www.hi-tm.com/Documents/Irradiation.html.

4. Marjorie Malley, "Bygone Spas: The Rise and Decay of Oklahoma's Radium Water," *The Chronicles of Oklahoma* 70, no. 4 (Winter 2002–2003), 446–67.

5. May Sybil Leslie to Arthur Smithells, Papers of Professor Arthur S. Smithells, 30 November 1909, Leeds University Library.

6. In addition to radioactivity, both Marie and Irène Curie were exposed to significant amounts of radiation from the mobile x-ray units they operated during World War I.

7. Otto Hahn, *My Life*, trans. Ernst Kaiser and Eithne Wilkins (London: MacDonald, 1970), 110.

Chapter 11

1. Frederick Soddy to Ernest Rutherford, 4 December 1903, Rutherford Papers, S116. Courtesy of the Syndics of Cambridge University Library. For more on the St. Joachimsthal mines, see Z. Zeman and P. Beneš, "St. Joachimsthal Mines and Their Importance in the Early History of Radioactivity," *Radiochimica Acta* 70/71 (1995): 23–29.

2. Soraya Boudia, *Marie Curie et son laboratoire* (Paris: Éditions des archives contemporaines, 2001), 116.

3. Ferdinand Heinrich, *Chemie und chemische Technologie radioaktiver Stoffe* (Berlin: Julius Springer, 1918), 288; United States Bureau of Mines, 1912, quoted in Edward R. Landa, *Buried Treasure to Buried Waste: The Rise and Fall of the Radium Industry* (Golden: Colorado School of Mines, 1987), 54.

4. On 2 August 1908 Kupelweiser wrote to the president of the Viennese Academy of Sciences: "I wish, so far as it is within my power, to prevent my fatherland from being disgraced by allowing its duty, given by Nature as a privilege, to be torn away by others ..." Stefan Meyer, "Das erste Jahrzehnt des Wiener Instituts für Radiumforschung," *Jahrbuch der Radioaktivität und Elektronik* 17 (1920): 1–29, on 2. Author's translation.

Chapter 12

1. Robert Reid, *Marie Curie* (New York: New American Library, 1974), 99.

2. Elizabeth Rona, *How It Came About* (Oak Ridge, TN: Oak Ridge Associated Universities, 1978), 15.

3. Stefan Meyer and Egon von Schweidler, *Radioaktivität* (Leipzig: B. G. Teubner, 1927), 497.

4. Otto Hahn, "Stefan Meyer," *Zeitschrift für Naturforschung* 5a (July 1950): 407–8.

5. Winifred Beilby Soddy was active in the suffragette movement. Among Soddy's students were Ruth Pirret and Ada Hitchens, who became his research assistant in 1922.

6. Robert W. Lawson, "The Part Played by Different Countries in the Development of the Science of Radioactivity," *Scientia* 30 (1921): 257–70.

7. Stefan Meyer to Ernest Rutherford, 22 January 1920, cited in Arthur S. Eve, *Rutherford* (New York: Macmillan, 1939), 278.

8. Stefan Meyer, "Lord Rutherford of Nelson," *Akademie der Wissenschaften zu Wien, Almanach* 88 (1938): 251–62, on 256–7; Arthur S. Eve, *Rutherford*, 243. Lawson published many papers in *Akademie der Wissenschaften zu Wien, Sitzungsberichte* during 1915–1918.

9. James Chadwick to Ernest Rutherford, 24 May 1918, Rutherford Papers, C24. Courtesy of the Syndics of Cambridge University Library. See also Chadwick's letters of 14 September 1915 (C22) and 31 March 1917 (C23).

10. Stefan Meyer to Ernest Rutherford, 22 January 1920, cited in Eve, *Rutherford*, 277; Rona, *How It Came About*, 26.

11. Ruth Lewin Sime, *Lise Meitner. A Life in Science* (Berkeley: University of California Press, 1996), 99; George von Hevesy, "A Scientific Career," *Perspectives in Biology and Medicine* 1 (Summer 1958): 345–65, on 353.

12. Otto Hahn, *My life*, trans. Ernst Kaiser and Eithne Wilkins (London: MacDonald, 1970), 136.

Chapter 13

1. Marjorie Malley, "Thermodynamics and Cold Light," *Annals of Science* 51 (1994): 203–24.

2. Frederick Soddy, "Radioactivity," *Annual Reports of the Chemical Society of London* 1 (1904): 244–80, on 279.

3. Albert Einstein, "Le principe de relativité et ses conséquences dans la physique moderne," *Archives des sciences physiques et naturelles* 29 (1910), 5–28, 125–44, on 144. Einstein's theory of mass-energy equivalence was not confirmed until the 1930s, when more information was available about atomic building blocks.

Chapter 14

1. Definition of "romantic" from *Webster's Third New International Dictionary*, unabridged.

2. Marie Curie, *Pierre Curie*, trans. Charlotte and Vernon Kellog (New York: Dover, 1963), 92; Eve Curie, *Marie Curie*, trans. Vincent Sheean (Garden City, NY: Doubleday, 1935), 341.

3. Marjorie Malley, "Bygone Spas: The Rise and Decay of Oklahoma's Radium Water," *Chronicles of Oklahoma* 80 (2002–2003), 446–67, on 461.
4. *Bhagavad-Gita*, Chapter 11:32. See Richard Rhodes, *The Making of the Atomic Bomb* (New York: Simon and Schuster, 1986), 676.

Sources for Epigraphs

Part I	Shakespeare, *Hamlet* Act i, Scene 5.
Part II	Eve Curie, *Marie Curie*, trans. Vincent Sheean (Garden City, NY: Doubleday, Doran & Company, 1937), 190.
Part III	"Vision of God's Creation," *Time*, November 3, 1975.
Chapter 1	W. Robert Nitske, *The Life of Wilhelm Conrad Röntgen Discoverer of the X Ray* (Tuscon: University of Arizona Press, 1971), 124.
Chapter 2	Eve Curie, *Madame Curie*, trans. Vincent Sheean (Garden City, NY: Doubleday, Doran & Company, 1937), 133; ibid, 191.
Chapter 3	Muriel Howarth, *Atomic Transmutation. The Greatest Discovery ever Made* (London: New World Publications, 1953), 56; ibid, 123; John B. Birks, *Rutherford at Manchester* (New York: W. A. Benjamin, 1963), 363.
Chapter 4	Arthur S. Eve, *Rutherford* (New York and Cambridge: Macmillan and Cambridge University Press, 1939), 107; Victor F. Hess, "Über Beobachtungen der durchdringenden Strahlung bei sieben Freiballonfahrten," *Physikalische Zeitschrift* 13 (1912), 1084–91, on 1090.
Chapter 5	Ernest Rutherford, *Radioactive Substances and their Radiations* (Cambridge: Cambridge University Press, 1913), 420; Muriel Howarth, *Atomic Transmutation, The Greatest Discovery Ever Made* (London: New World Publications, 1953), 64.
Chapter 6	Frederick Soddy, "The Chemistry of Mesothorium," *Journal of the Chemical Society of London* 99 (1911).
Chapter 7	Henri Poincaré, *Dernières Pensées*, ed. Ernest Flammarion (Paris: Ernest Flammarion, 1919, originally pub. 1913), 204; Henry G. J. Moseley, "The High-Frequency Spectra of the Elements," *Philosophical Magazine* 26 (1913), 1024–34, on 1031.
Chapter 8	Arthur S. Eve, *Rutherford* (New York and Cambridge: Macmillan and Cambridge University Press, 1939), 224.
Chapter 9	Horace, *Satires*, Book I-1, line 106.
Chapter 10	Otto Hahn, *My Life*, trans. Ernst Kaiser and Eithne Wilkins (London: MacDonald, 1970), 110.
Chapter 11	Horace, *Odes*, Book II:xvi, line 27.

Chapter 12	*McClure's Magazine*, April 1896, cited in W. Robert Nitske, *The Life of Wilhelm Conrad Röntgen Discoverer of the X Ray* (Tuscon: University of Arizona Press, 1971), 130; Marie Curie, *Pierre Curie*, trans. by Charlotte and Vernon Kellog (New York: Dover, 1963, originally published 1923), 70.
Chapter 13	*Ecclesiastes* 1,9.
Chapter 14	Albert Einstein, *Living Philosophies* (New York: Simon & Schuster, 1931), 6.

SELECTED BIBLIOGRAPHY

Scientific publications and archival materials

Comprehensive references to original publications and archival sources for part I and sections of parts II and III are in Marjorie Malley, *From Hyperphosphorescence to Nuclear Decay: A History of the Early Years of Radioactivity, 1896–1914* (Ann Arbor, MI: University Microfilms, 1976).

Important sources for radioactivity's scientific literature are Stefan Meyer and Egon von Schweidler, *Radioaktivität* (Leipzig: B. G. Teubner, 1916, 1927); Ernest Rutherford, *Radio-activity* (Cambridge: Cambridge University Press, 1904, 1905), and *Radioactive Substances and Their Radiations* (Cambridge: Cambridge University Press, 1913); Marie Curie, *Traité de radioactivité* (Paris: Gauthier-Villars, 1910); and *Jahrbuch der Radioaktivität und Elektronik* (Leipzig: 1904–1920).

Other resources and further reading

This bibliography primarily contains publications in English. A few works in French and German are included.

PART I

Badash, Lawrence. *Radioactivity in America*. Baltimore: Johns Hopkins University Press, 1979.

Birks, John B., ed. *Rutherford at Manchester*. New York: W. A. Benjamin, 1963.

Brian, Denis. *The Curies. A Biography of the Most Controversial Family in Science*. Hoboken, NJ: John Wiley and Sons, 2005.

Broda, Engelbert. *Ludwig Boltzmann*. Translated by Larry Gay and the author. Woodbridge, CT: Ox Bow Press, 1983.

Brown, George I. *Invisible Rays. The History of Radioactivity*. Stroud, UK: Sutton, 2002.

Brush, Stephen G. *Statistical Physics and the Atomic Theory of Matter*. Princeton, NJ: Princeton University Press, 1983.

Buchwald, Jed Z., and Andrew Warwick. *Histories of the Electron: The Birth of Microphysics*. Cambridge, MA: MIT Press, 2001.

Burchfield, Joe D. *Lord Kelvin and the Age of the Earth*. New York: Science History Publications, 1975.

Burnet, John. *Early Greek Philosophy*. New York: Meridan Books, 1957. First published 1920.

Coen, Deborah R. *Vienna in the Age of Uncertainty*. Chicago: University of Chicago Press, 2007.

Curie, Eve. *Madame Curie*. Translated by Vincent Sheean. Garden City, NY: Doubleday, Doran & Company, 1937.

Curie, Marie. *Pierre Curie*. Translated by Charlotte and Vernon Kellog. New York: Dover, 1963. First published 1923.

Dahl, Per F. *Flash of the Cathode Rays*. Bristol, UK: Institute of Physics Publishing, 1997.

DeKosky, Robert. "Spectroscopy and the Elements in the Late Nineteenth Century: The Work of Sir William Crookes." *British Journal for the History of Science* 6 (1972–73): 400–23.

Duhem, Pierre. *The Aim and Structure of Physical Theory*. Translated by Philip P. Wiener. New York: Atheneum, 1962. First published 1914.

Eve, Arthur S. *Rutherford*. New York: Macmillan, 1939.

Fricke, Rudolf G. A. *Friedrich Oscar Giesel*. Wolfenbüttel: AF Verlag, 2001.

———. *J. Elster & H. Geitel*. Braunschweig: Döring Druck, Druckerei und Verlag, 1992.

Gigerenzer, Gerd, Zeno Swijtink, Theodore Porter, Lorraine Daston, John Beatty, and Lorenz Krüger. *The Empire of Chance*. Cambridge: Cambridge University Press, 1989.

Glasser, Otto. *Wilhelm Conrad Röntgen and the Early History of the Roentgen Rays*. Springfield, IL: Charles C. Thomas, 1934.

Hahn, Otto. *A Scientific Autobiography*. Translated and edited by Willy Ley. New York: Scribner's Sons, 1966.

———. *My Life*. Translated by Ernst Kaiser and Eithne Wilkins. London: MacDonald, 1970.

Heilbron, John, ed. *The Oxford Companion to the History of Modern Science*. New York: Oxford University Press, 2003.

———. "The Scattering of α and β Particles and Rutherford's Atom." *Archives for History of Exact Sciences* 4 (1968): 247–307.

———. *H. G. J. Moseley: The Life and Letters of an English Physicist, 1887–1915*. Berkeley: University of California Press, 1974.

Hoffmann, Dieter, Fabio Bevilacqua, and Roger H. Stuewer, eds. *The Emergence of Modern Physics*. Pavia, Italy: La Goliardica Pavese, 1996.

Howarth, Muriel. *Atomic Transmutation. The Greatest Discovery Ever Made*. London: New World Publications, 1953.

————. *Pioneer Research on the Atom. The Life Story of Frederick Soddy.* London: New World Publications, 1953.

Jensen, Carsten. *Controversy and Consensus: Nuclear Beta Decay 1911–1934.* Basel: Birkhäuser, 2000.

Kauffman, George B., ed. *Frederick Soddy (1877–1956): Early Pioneer in Radiochemistry.* Boston: D. Reidel, 1986.

Kevles, Bettyann Holtzmann. *Naked to the Bone. Medical Imaging in the Twentieth Century.* Brunswick, NJ: Rutgers University Press, 1997.

Kirby, Harold W. *The Early History of Radiochemistry.* Atomic Energy Commission Research and Development Report. Miamisburg, OH: Monsanto Research Corp., Mound Laboratory, 1972.

Kragh, Helge. *Quantum Generations: A History of Physics in the Twentieth Century.* Princeton, NJ: Princeton University Press, 2002.

Malley, Marjorie. "The Discovery of Atomic Transmutation: Scientific Styles and Philosophies in France and Britain." *Isis* 70 (1979): 213–23. Reprinted in *History of Physics. Selected Reprints.* Edited by Stephen G. Brush, 184–94. College Park, MD: American Association of Physics Teachers, 1988.

————. "The Discovery of the Beta Particle." *American Journal of Physics* 39 (1971): 1454–61. Reprinted in *Physics History from AAPT Journals.* Edited by Melba N. Phillips, 83–90. College Park, MD: American Association of Physics Teachers, 1985.

Martins, Roberto De Andrade. "Becquerel and the Choice of Uranium Compounds." *Archive for History of Exact Sciences* 51 (1997): 67–81.

Merricks, Linda. *The World Made New: Frederick Soddy, Science, Politics, and Environment.* Oxford: Oxford University Press, 1996.

Nitske, W. Robert. *The Life of Wilhelm Conrad Röntgen Discoverer of the X Ray.* Tucson: University of Arizona Press, 1971.

Nobel Lectures. Chemistry, 1901–1921. Amsterdam: Elsevier, 1966.

Nobel Lectures. Physics, 1901–1921. Amsterdam: Elsevier, 1967.

Nye, Mary Jo. "Gustave Le Bon's Black Light: A Study of Physics and Philosophy in France at the Turn of the Century." *Historical Studies in the Physical Sciences* 4 (1974): 163–96.

————. *The Cambridge History of Science.* Vol. 5, The Modern Physical and Mathematical Sciences. Cambridge: Cambridge University Press, 2003.

Pais, Abraham. *Inward Bound. Of Matter and Forces in the Physical World.* Oxford: Clarendon Press and Oxford University Press, 1986.

Poincaré, Henri. *Science and Hypothesis.* Translated by W. J. G. [sic]. New York: Dover, 1952 First published 1905.

Quinn, Susan. *Marie Curie. A Life.* New York: Simon & Schuster, 1995.

Reid, Robert. *Marie Curie.* New York: New American Library, 1974.

Romer, Alfred. *The Discovery of Radioactivity and Transmutation.* New York: Dover, 1964.

————. *Radiochemistry and the Discovery of Isotopes*. New York: Dover, 1970.

Shea, William R., ed. *Otto Hahn and the Rise of Nuclear Physics*. Dordrecht: D. Reidel, 1983.

Sinclair, S. B. "Crookes and Radioactivity: From Inorganic Evolution to Atomic Transmutation." *Ambix* 32 (1985): 15–31.

Stuewer, Roger, ed. *Nuclear Physics in Retrospect*. Minneapolis: University of Minnesota Press, 1979.

Trenn, Thaddeus. *The Self-Splitting Atom*. London: Taylor & Francis, 1977.

Weeks, Mary E. *Discovery of the Elements*. Easton, PA: Journal of Chemical Education, 1968.

Wilson, David. *Rutherford. Simple Genius*. Cambridge, MA: MIT Press, 1983.

PART II

Allison, Malorye. "The Radioactive Elixir." *Harvard Magazine* 94, no. 3 (1992): 73–75.

Berndt, G. *Radioaktive Leuchtfarben*. Braunschweig: Friedrich Vieweg, 1920.

Boudia, Soraya. *Marie Curie et son laboratoire*. Paris: Éditions des archives contemporaines, 2001.

————. "Marie Curie: Scientific Entrepreneur." *Physics World* 11 (1998): 35–39.

Brock, William H. *William Crookes (1832–1919) and the Commercialization of Science*. Aldershot, UK: Ashgate, 2008.

Caufield, Catherine. *Multiple Exposures. Chronicles of the Radiation Age*. Chicago: University of Chicago Press, 1989.

Clark, Claudia: *Radium Girls. Women and Industrial Health Reform, 1910–1935*. Chapel Hill: University of North Carolina Press, 1997.

Gibbon, Anthony. "Uranium from Oil Flood Waters." *World Oil* (May 1956): 62–63.

Heinrich, Ferdinand. *Chemie und chemische Technologie radioaktiver Stoffe*. Berlin: Julius Springer, 1918.

Hughes, Jeff. *The Manhattan Project*. New York: Columbia University Press, 2002.

Landa, Edward R. *Buried Treasure to Buried Waste: The Rise and Fall of the Radium Industry*. Golden: Colorado School of Mines, 1987.

————. "The First Nuclear Industry." *Scientific American* 247 (1982): 180–193.

Malley, Marjorie. "Bygone Spas: The Rise and Decay of Oklahoma's Radium Water." *Chronicles of Oklahoma* 80 (2002–2003): 446–67.

McCoy, Roger M., ed. *Radiometric Surveys in Petroleum Exploration*. Denver: Association of Petroleum Geochemical Explorationists, 1995.

Radium. Brussels: Radium Belge (Union Minière du Haut Katanga), 1925.

Silverman, Alexander. "Pittsburgh's Contribution to Radium Recovery." *Journal of Chemical Education* 27 (1950): 303–8.

Walker, J. Samuel. *Permissible Dose*. A History of Radiation Protection in the Twentieth Century. Berkeley: University of California Press, 2000.

Zoellner, Tom. "The Uranium Rush." *Invention & Technology* 16 (Summer 2000): 57–63.

PART III

Badash, Lawrence. "British and American Views of the German Menace during World War I." *Notes and Records of the Royal Society of London* 34, July 1979: 91–121.

Cahan, David. *An Institute for an Empire. The Physikalisch-Technische Reichanstalt 1871–1918*. Cambridge: Cambridge University Press, 1989.

Crawford, Elisabeth. *Nationalism and Internationalism in Science, 1880–1939*. Cambridge: Cambridge University Press, 1992.

Davis, J. L. "The Research School of Marie Curie in the Paris Faculty, 1907–14." *Annals of Science* 52 (1995): 321–55.

Geison, Gerald L. and Frederic L. Holmes, eds. Research Schools. Historical Reappraisals. *Osiris* 8 (1993).

Grinstein, Louise S., Rose K. Rose, and Miriam H. Rafailovich, eds. *Women in Chemistry and Physics*. Westport, CT: Greenwood Press, 1993.

Hanle, Paul A. "Indeterminacy before Heisenberg: The Case of Franz Exner and Erwin Schrödinger." *Historical Studies in the Physical Sciences* 10 (1979): 225–69.

Holton, Gerald. *The Scientific Imagination: Case Studies*. Cambridge: Cambridge University Press, 1978.

Malley, Marjorie. "Thermodynamics and Cold Light." *Annals of Science* 51 (1994): 203–22.

Merz, John Theodore. *A History of European Thought in the Nineteenth Century*. Vols. 1 and 2. New York: Dover, 1965. First published c. 1904.

Nye, Mary Jo. *Before Big Science*. New York: Twayne Publishers, 1996.

Paul, Harry W. *The Sorcerer's Apprentice: The French Scientist's Image of German Science, 1840–1919*. Gainsville: University of Florida Press, 1972.

Pycior, Helena M., Nancy M. Slack, and Pnina G. Abir-Am, eds. *Creative Couples in the Sciences*. New Brunswick, NJ: Rutgers University Press, 1996.

Rayner-Canham, Marelene and Geoffrey W. Rayner-Canham. *Harriet Brooks, Pioneer Nuclear Scientist*. Montréal: McGill-Queen's University Press, 1992.

———. *A Devotion to Their Science. Pioneer Women of Radioactivity*. Philadelphia: Chemical Heritage Foundation, 1997.

Rentetzi, Maria. "Gender, Politics, and Radioactivity Research in Interwar Vienna." *Isis* 95 (2004): 359–93.

———. *Trafficking Material and Gendered Experimental Practices: Radium Research in Early 20th Century Vienna*. New York: Columbia University Press, 2008.

Rhodes, Richard. *The Making of the Atomic Bomb*. New York: Simon and Schuster, 1986.

Rona, Elizabeth. *How It Came About*. Oak Ridge, TN: Oak Ridge Associated Universities, 1978.

————. "Laboratory Contamination in the Early Period of Radiation Research." *Health Physics* 37 (December 1979): 723–27.

Schroeder-Gudehus, Brigitte. "Nationalism and Internationalism." In *Companion to the History of Modern Science*, edited by R. C. Colby, G. N. Cantor, J. R. R. Christie, and M. J. S. Hodge, 909–19. London: Routledge, 1990.

Siegel, Daniel M. "Classical-Electromagnetic and Relativistic Approaches to the Problem of Nonintegral Atomic Masses." *Historical Studies in the Physical Sciences* 9 (1978): 323–60.

Sime, Ruth Lewin. *Lise Meitner. A Life in Science*. Berkeley: University of California Press, 1996.

Weart, Spencer. *Nuclear Fear. A History of Images*. Cambridge, MA: Harvard University Press, 1988.

REFERENCE WORKS

Funk & Wagnalls Standard Dictionary of Folklore, Mythology, and Legend. Edited by Maria Leach. New York: Funk & Wagnalls Company, 1949, 1950; New York: Funk and Wagnalls Publishing Company, 1972.

Grun, Bernard. *The Timetables of History*. New York: Simon & Schuster, 1991.

Hellmans, Alexander, and Bryan Bunch. *The Timetables of Science*. New York: Simon and Schuster, 1988.

Ogilvie, Marilyn, and Joy Harvey, eds. *The Biographical Dictionary of Women in Science*. New York: Routledge, 2000.

Oxford Dictionary of Quotations. Oxford: Oxford University Press, 1980.

Simpson, James B., comp. *Simpson's Contemporary Quotations*. Boston: Houghton Mifflin, 1988.

INDEX OF PERSONS

INDEX OF SUBJECTS